Education in a Competitive and Globalizing World

# Lifelong Learning: Theoretical and Practical Perspectives on Adult Numeracy and Vocational Mathematics

# EDUCATION IN A COMPETITIVE AND GLOBALIZING WORLD

**Success in Mathematics Education**
*Caroline B. Baumann (Editor)*
2009. 978-1-60692-299-6

**Mentoring: Program Development, Relationships and Outcomes**
*Michael I. Keel (Editor)*
2009. 978-1-60692-287-3

**Motivation in Education**
*Desmond H. Elsworth (Editor)*
2009. 978-1-60692-234-7

**Evaluating Online Learning: Challenges and Strategies for Success**
*Arthur T. Weston (Editor)*
2009. 978-1-60741-107-9

**Enhancing Prospects of Longer-Term Sustainability of Cross-Cultural INSET Initiatives in China**
*Chunmei Yan (Author)*
2009. 978-1-60741-615-9

**Reading at Risk: A Survey of Literary Reading in America**
*Rainer D. Ivanov (Author)*
2009. 978-1-60692-582-9

**Reading: Assessment, Comprehension and Teaching**
*Nancy H. Salas and Donna D. Peyton (Editors)*
2009. 978-1-60692-615-4

**Multimedia in Education and Special Education**
*Onan Demir and Cari Celik (Editors)*
2009. 978-1-60741-073-7

**Rural Education in the 21$^{st}$ Century**
*Christine M.E. Frisiras (Editors)*
2009. 978-1-60692-966-7

**Nutrition Education and Change**
*Beatra F. Realine (Editor)*
2009. 978-1-60692-983-4

**What Content-Area Teachers Should Know About Adolescent Literacy**
*National Institute for Literacy (Author)*
2009. 978-1-60741-137-6

**The Reading Literacy of U.S. Fourth-Grade Students in an International Context**
*Justin Baer, Stéphane Baldi, Kaylin Ayotte, Patricia J. Green and Daniel McGrath (Authors)*
2009. 978-1-60741-138-3

**Teacher Qualifications and Kindergartners' Achievements**
*Cassandra M. Guarino, Laura S. Hamilton, J.R. Lockwood, Amy H. Rathbun and Elvira Germino Hausken (Authors)*
2009. 978-1-60741-180-2

**PCK and Teaching Innovations**
*Syh-Jong Jang (Author)*
2009. 978-1-60741-147-5

**IT- Based Project Change Management System**
*Faisal Manzoor Arain and Low Sui Pheng (Authors)*
2009. 978-1-60741-148-2

**Learning in the Network Society and the Digitized School**
*Rune Krumsvik (Editor)*
2009. 978-1-60741-172-7

**Effects of Family Literacy Interventions on Children's Acquisition of Reading**
*Ana Carolina Pena (Editor)*
2009. 978-1-60741-236-6

**Developments in Higher Education**
*Mary Lee Albertson (Editor)*
2009. 978-1-60876-113-5

**Career Development**
*Hjalmar Ohlsson and Hanne Borg (Editors)*
2009. 978-1-60741-464-3

**Special Education in the 21$^{st}$ Century**
*MaryAnn T. Burton (Editor)*
2009. 978-1-60741-556-5

**Academic Self-Beliefs in the Spelling Domain:**
*Günter Faber (Author)*
2009. 978-1-60741-603-6

**Approaches to Early Childhood and Elementary Education**
*Francis Wardle (Author)*
2009. 978-1-60741-643-2

**Challenges of Quality Education in Sub-Saharan African Countries**
*Daniel Namusonge Sifuna and Nobuhide Sawamura (Authors)*
2009. 978-1-60741-509-1

**Academic Administration: A Quest for Better Management and Leadership in Higher Education**
*Sheying Chen (Editor)*
2009. 978-1-60741-732-3

**Recent Trends in Education**
*Borislav Kuzmanoviæ and Adelina Cuevas (Editors)*
2009. 978-1-60741-795-8

**Expanding Teaching and Learning Horizons in Economic Education**
*Franklin G. Mixon, Jr. and Richard J. Cebula (Editors)*
2009. 978-1-60741-971-6

**New Research in Education: Adult, Medical and Vocational**
*Edmondo Balistrieri and Giustino DeNino (Editors)*
2009. 978-1-60741-873-3

**Collaborative Learning: Methodology, Types of Interactions and Techniques**
*Edda Luzzatto and Giordano DiMarco (Editors)*
2009. 978-1-60876-076-3

**Handbook of Lifelong Learning Developments**
*Margaret P. Caltone (Editor)*
2009. 978-1-60876-177-7

**Reading: Assessment, Comprehension and Teaching**
*Nancy H. Salas and Donna D. Peyton (Editors)*
2009. 978-1-60876-543-0

**Mentoring: Program Development, Relationships and Outcomes**
*Michael I. Keel (Editor)*
2009. 978-1-60876-727-4

**Academic Administration: A Quest for Better Management and Leadership in Higher Education**
*Sheying Chen (Editor)*
2009. 978-1-61668-571-3

**Sexuality Education**
*Kelly N. Stanton (Author)*
2010. 978-1-60692-153-1

**Disadvantaged Students and Crisis in Faith-Based Urban Schools**
*Thomas G. Wilson (Author)*
2010. 978-1-60741-535-0

**Educational Change**
*Aden D. Henshall and Bruce C. Fontanez (Editors)*
2010. 978-1-60876-389-4

**Delving into Diversity: An International Exploration of Issues of Diversity in Education**
*Vanessa Green and Sue Cherrington (Editors)*
2010. 978-1-60876-361-0

**Health Education: Challenges, Issues and Impact**
*André Fortier and Sophie Turcotte (Editors)*
2010. 978-1-60876-568-3

**Educational Games: Design, Learning and Applications**
*Frej Edvardsen and Halsten Kulle (Editors)*
2010. 978-1-60876-692-5

**Attracting International Students for Higher Education**
*Russell C. Carswell (Editor)*
2010. 978-1-60741-666-1

**The Process of Change in Education: Moving from Descriptive to Prescriptive Research**
*Baruch Offir (Author)*
2010. 978-1-60741-451-3

**Adopting Blended Learning for Collaborative Work in Higher Education**
*Alan Hogarth (Author)*
2010. 978-1-60876-260-6

**America's Historically Black Colleges and Universities**
*Giovani Lucisano (Editor)*
2010. 978-1-60741-510-7

**Medical Education: The State of the Art**
*Rossana Salerno-Kennedy and Siún O'Flynn (Editors)*
2010. 978-1-60876-194-4

**Reading in 2010; A Comprehensive Review of a Changing Field**
*Michael F. Shaughnessy (Editor)*
2010. 978-1-60876-659-8

**Music Education**
*João Hermida and Mariana Ferreo (Editors)*
2010. 978-1-60876-655-0

**Computer-Assisted Teaching: New Developments**
*Brayden A. Morris and George M. Ferguson (Editors)*
2010. 978-1-60876-855-4

**Handbook of Curriculum Development**
*Limon E. Kattington (Editor)*
2010. 978-1-60876-527-0

**Sexuality Education and Attitudes**
*Jovan Stanovic and Milo Lalic (Editors)*
2010. 978-1-60741-662-3

**National Financial Literacy Strategy**
*Toma P. Hendriks (Editor)*
2010. 978-1-60741-827-6

**Virtual Worlds: Controversies at the Frontier of Education**
*Kieron Sheehy, Rebecca Ferguson and Gill Clough (Editors)*
2010. 978-1-60876-261-3

**Education in Asia**
*Hiroto S. Nakamura (Editor)*
2010. 978-1-61668-617-8

**Becoming an Innovative Teacher Educator: Designing and Developing a Successful Hybrid Course**
*Quiyun Lin (Author)*
2010. 978-1-60876-465-5

**Higher Education: Teaching, Internationalization and Student Issues**
*Magnus E. Poulsen (Editor)*
2010. 978-1-61668-548-5

**Educational Assessment: Measuring Progress in Schools**
*Oliver T. Ricoha (Editor)*
2010. 978-1-60876-818-9

**Decentralization and Marketization: The Changing Landscape of China's Adult and Continuing Education**
*Ning Rong Liu (Author)*
2010. 978-1-60876-835-6

**Issues in English Language Learning**
*Gerald. E. Albertson (Editor)*
2010. 978-1-60876-848-6

**Multiculturalism in the Age of the Mosaic**
*Michael Ọladẹjọ Afọláyan (Editor)*
2010. 978-1-60876-995-7

**Reading First Impact Study - Part II**
*Martha O. Parkis (Editor)*
2010. 978-1-60876-891-2

**Collaborative and Individual Learning in Teaching**
*Julien Mercie, Caroline Girard, Monique Brodeur and Line Laplante (Authors)*
2010. 978-1-60876-889-9

**Postsecondary Education and Student Aid**
*Jennifer G. Hartley (Editor)*
2010. 978-1-60876-935-3

**Education to Meet New Challenges in a Networked Society**
*Leo Jansen, Paul Weaver, Rietje van Dam-Mieras (Authors)*
2010. 978-1-61668-245-3

**Time Out: Examining Seclusion and Restraint in Schools**
*Laura E. Kentley (Editor)*
2010. 978-1-60876-932-2

**Education Reforms in Ghana: Curriculum in Junior High Schools**
*AGeorge M. Osei (Athors)*
2010. 978-1-60876-948-3

**Effective Teaching for Intended Learning Outcomes in Science and Technology (Metilost)**
*Bernardino Lopes, Paulo Cravino and Silva (Authors)*
2010. 978-1-60876-958-2

**Specialized Rasch Measures Applied at the Forefront of Education**
*Russell F. Waugh(Editor)*
2010. 978-1-61668-032-9

**Applications of Rasch Measurement in Education**
*Russell Waugh (Editor)*
2010. 978-1-61668-026-8

**Striving for the Perfect Classroom: Instructional and Assessment Strategies to Meet the Needs of Today's Diverse Learners**
*Kelli R. Paquette and Sue A. Rieg (Authors)*
2010. 978-1-61668-039-8

**Medical Education in the New Millennium**
*Amal A. El-Moamly (Author)*
2010. 978-1-61668-456-3

**Medical Education in the New Millennium**
*Amal A. El-Moamly (Author)*
2010. 978-1-61668-206-4

**Lifelong Learning: Theoretical and Practical Perspectives on Adult Numeracy and Vocational Mathematics**
*Gail E. FitzSimons (Author)*
2010. 978-1-61668-291-0

**Improving Visual Teaching Materials**
*Adrianne Rourke and Zena O'Connor (Authors)*
2010. 978-1-61668-294-1

**Handbook of Curriculum Development**
*Limon E. Kattington (Editor)*
2010. 978-1-61668-350-4

**Attracting International Students for Higher Education**
*Russell C. Carswell (Editor)*
2010. 978-1-61668-388-7

**Postsecondary Education and Student Aid**
*Jennifer G. Hartley (Editor)*
2010. 978-1-61668-428-0

**Spelling Skills: Acquisition, Abilities, and Reading Connection**
*Blake C. Fabini (Editor)*
2010. 978-1-61668-472-3

**Kindergartens: Programs, Functions and Outcomes**
*Spencer B. Thompson (Editor)*
2010. 978-1-61668-530-0

**Spelling Skills: Acquisition, Abilities, and Reading Connection**
*Blake C. Fabini (Editor)*
2010. 978-1-61668-531-7

**Improving Visual Teaching Materials**
*Adrianne Rourke and Zena O'Connor (Authors)*
2010. 978-1-61668-700-7

**Lifelong Learning: Theoretical and Practical Perspectives on Adult Numeracy and Vocational Mathematics**
*Gail E. FitzSimons (Author)*
2010. 978-1-61668-705-2

**Kindergartens: Programs, Functions and Outcomes**
*Spencer B. Thompson (Editor)*
2010. 978-1-61668-711-3

**Education in a Competitive and Globalizing World**

# LIFELONG LEARNING: THEORETICAL AND PRACTICAL PERSPECTIVES ON ADULT NUMERACY AND VOCATIONAL MATHEMATICS

## GAIL E. FITZSIMONS

Nova Science Publishers, Inc.
*New York*

Copyright © 2010 by Nova Science Publishers, Inc.

**All rights reserved.** No part of this book may be reproduced, stored in a retrieval system or transmitted in any form or by any means: electronic, electrostatic, magnetic, tape, mechanical photocopying, recording or otherwise without the written permission of the Publisher.

For permission to use material from this book please contact us:
Telephone 631-231-7269; Fax 631-231-8175
Web Site: http://www.novapublishers.com

**NOTICE TO THE READER**

The Publisher has taken reasonable care in the preparation of this book, but makes no expressed or implied warranty of any kind and assumes no responsibility for any errors or omissions. No liability is assumed for incidental or consequential damages in connection with or arising out of information contained in this book. The Publisher shall not be liable for any special, consequential, or exemplary damages resulting, in whole or in part, from the readers' use of, or reliance upon, this material.

Independent verification should be sought for any data, advice or recommendations contained in this book. In addition, no responsibility is assumed by the publisher for any injury and/or damage to persons or property arising from any methods, products, instructions, ideas or otherwise contained in this publication.

This publication is designed to provide accurate and authoritative information with regard to the subject matter covered herein. It is sold with the clear understanding that the Publisher is not engaged in rendering legal or any other professional services. If legal or any other expert assistance is required, the services of a competent person should be sought. FROM A DECLARATION OF PARTICIPANTS JOINTLY ADOPTED BY A COMMITTEE OF THE AMERICAN BAR ASSOCIATION AND A COMMITTEE OF PUBLISHERS.

LIBRARY OF CONGRESS CATALOGING-IN-PUBLICATION DATA

Available upon Request
ISBN: 978-1-61668-291-0

*Published by Nova Science Publishers, Inc.* ✢ *New York*

# CONTENTS

| | | |
|---|---|---|
| **Preface** | | xvii |
| **Chapter 1** | Introduction | 1 |
| **Chapter 2** | Five Paradigms of Education | 5 |
| **Chapter 3** | Distinctions between Mathematics and Numeracy | 15 |
| **Chapter 4** | Perspectives on Activity Theory | 23 |
| **Chapter 5** | Activity Theory and Technology in Education | 31 |
| **Chapter 6** | Implications of an Activity Theory Framework for Lifelong Learning in Mathematics and Numeracy | 37 |
| **Chapter 7** | A Framework for Planning and Evaluating Formal Mathematics and Numeracy Lifelong Learning Programmes | 43 |
| **Chapter 8** | Conclusion | 51 |
| **References** | | 55 |
| **Index** | | 61 |

# PREFACE

For a variety of historical, cultural, social, and/or economic reasons adults may experience the need to continue their mathematics education in some form. The concept of lifelong learning, posited by Dewey in 1916, has been widely recognised since the 1970s. However, this concept is contested and there are many perspectives which may be in tension or even contradiction according to the lens adopted. The chapter will review some of these perspectives because the dominant one will affect the educational strategies and outcomes.

The concept of adult numeracy — which arose in Britain in the 1950s — is becoming increasingly common and is also contested, as is the terminology (e.g., quantitative literacy). Drawing on the work of Bernstein, I will distinguish between mathematics and numeracy as the choice made can also have educational implications, although I have argued elsewhere for a convergence.

For mathematics in particular, the concept of transfer is a vexed issue. It is commonly assumed that what is learned in the formal classroom or other learning site will automatically be recognised and applied unproblematically in a very different situation such as the workplace. Curriculum and pedagogy for adult numeracy and vocational mathematics need to take cognisance of this and other important issues to avoid the situation of merely replicating the kinds of 'school' mathematics that many people, young and old, have been failed by and continue to avoid engaging with.

In today's world, technology is playing an increasingly important role in educational situations, in the workplace, and at home. Technology plays a dual role in the teaching and learning of mathematics/numeracy for adults. Technology, electronic and otherwise, offers a medium to enhance learning in the form of tools such as rulers and compasses as well as software programs. Technology in the form of calculators of various kinds or computer applications, such as spreadsheets or statistics programs can act as aids to overcome the limitations of human memory or to support the exploration of new ideas by experimentation or simulation. However, these need to be made objects of learning in their own right before they can support higher level thinking. Electronic technologies offer increasingly sophisticated means of communication, in education and at work or society at large.

*Chapter 1*

# INTRODUCTION

The concept of lifelong learning, posited by Dewey in 1916, has been widely recognised since the 1970s (e.g., Delors, 1996, EU, 1996a, 1996b, 2000; OECD, 1996; UIL, 2008; UNESCO, 2000, 2005,; UNESCO & ILO, 2002). In their philosophical discussion, Aspin and Chapman (2001) identify the shift in terminology from "lifelong education" to "lifelong learning." They prefer the latter because of its value neutrality and the focus on the individual rather than the institution in the construction of programmes of activities where processes — ongoing activities — are emphasised over product. From an epistemological perspective, they claim that lifelong learners acquire additional beliefs, knowledge and understanding within and outside of conventional education institutions, putting their hypotheses and tentative solutions to the test of criticism in arenas they find appropriate, traditional or non-traditional. Aspin and Chapman assert the "triadic" nature of lifelong learning as a complex interplay between the aims and undertakings of:

- economic progress and development,
- personal development and fulfilment, and
- social inclusiveness and democratic understanding and activity.

Lifelong learning requires a multi-faceted approach where the individual learners are able to make meaning on a personal level, requiring that they have "the skills of research, enquiry, and self-starting curiosity, that is constantly seeking answers to questions posed to them by others or by their own situations in life, their problems and predicaments" (p. 21). In other words, they need to learn how to learn.

Interestingly, the three aims and approaches to lifelong learning compare with those of Niss (1996) who listed fundamental reasons for the formal teaching of mathematics as including:

- contributing to the *technological and socio-economic development* of society at large, either as such or in competition with other societies/countries;
- contributing to society's *political, ideological and cultural maintenance and development*, again either as such or in competition with other societies/countries;
- providing *individuals with prerequisites which may help them to cope with life* in the various spheres in which they live: education or occupation; private life; social life; life as a citizen. (p. 13)

As Aspin and Chapman (2001) observe, it is easy to fall into an essentialist trap, in defining lifelong learning. Fixed immutable definitions are always risky as lifelong learning — in common with other spheres of education — is context-bound. In some senses lifelong learning encapsulates the normal human processes from cradle to grave. Formal education processes themselves vary according to time and place on a global level in terms of when, where, and how they take place, and for whom. Vocational education offers a classic example of contestability. For some, it is narrowly defined to be education of a technical nature preparing or retraining workers for manual-type occupations including the trades and service industries, often delivered in institutions separate from the mainstream of schooling or university education; sometimes before paid work can begin and sometimes concurrent with it. This view generally accords vocational education a lower status than higher school or university qualifications. For others (e.g. Maglen, 1996), much of university study can be regarded as vocational. Workplace education may be considered as learning on-the-job, off-the-job, formally or informally or both, and could potentially encompass the whole range of employees from unskilled or semi-skilled up to the highest levels of management. The focus of this book is on lifelong learning for people who are beyond the age of compulsory schooling and are in the paid workforce or engaged in unpaid labour, who may be formally enrolled in education institutions or other accredited sites of learning such as the workplace or a community setting, or who may be learning on an informal basis as part of their everyday activities in the workplace and elsewhere. It does not specifically address university education although the issues discussed may be of relevance. The object of learning is claimed to be numeracy — although, as discussed below, this is also a contested term — and vocational mathematics. Numeracy is not taken to be synonymous with mathematics, although there are clearly commonalities.

This book will begin by considering five broad paradigms of teaching and learning that have been in operation during the last century or so. This will set the scene for the kinds of formal education experiences that adults may have encountered in the past — for good or for bad — and may yet encounter should they return to formal study once more. Past experiences set up a range of expectations that cannot be ignored. For their part, practitioners should be aware of these and have the ability to draw on those aspects of the different paradigms in their repertoire that they consider valuable at the time. The third chapter draws upon the work of Basil Bernstein to differentiate numeracy and mathematics from an analytical perspective. The fourth chapter introduces activity theory and describes each element of Engeström's mediational triangle which offers a model of learning that combines a psychological focus on the individual learner in pursuit of a certain object assisted by tools or mediating artefacts, and supported by the social and cultural contexts of the broader community involvement, the rules and expectations of both the educational setting and of the discipline, and the division of labour. Taking activity theory to a finer grade of analysis, the fifth chapter focuses particularly on the use of technology in vocational mathematics and adult numeracy classes. The sixth chapter offers a framework for planning and evaluating adult numeracy/mathematics programs based on an approach related to Engeström's observations of activity theory at work. A brief concluding chapter ends the book.

# EDUCATION POLICY AND PARADIGMS OF TEACHING AND LEARNING

Education policy is always contested, from problem statement to formulation to interpretation through to implementation. Societal contexts and changing expectations, changes of government, or even changes of responsible politicians and bureaucrats result in major or minor shifts in intended outcomes for learners and the perceptions held by decision makers of the optimal strategies for achieving their goals. However, policies implemented on a grand scale for all learners in a given society will work well for some learners and have negative consequences for others. As a relatively privileged young woman who went from private school to university and straight into mathematics teaching in a secondary school, and who actually enjoyed much of her mathematics education, I was a person who was suited by the traditional system of education of my time. As an adult educator, I have since come into contact with many learners of all ages who were not. I have also known many adults who might have been well suited but who, for a variety of social, cultural, economic, and historic reasons, were unable to pursue the mathematics (and other) education they desired in their school years. In developed countries at least, adults now have opportunities to continue their formal education — albeit sometimes with a degree of coercion. There are also often expectations, explicit or implicit, that workers will continue their learning throughout their working lives, on-the-job, and possibly through formal institutions.

When adults are engaged in lifelong learning activities their previous experiences of formal education, however long or short, are likely to have a formative influence on subsequent learning, especially their attitudes and expectations. Most people can identify positive and negative experiences and, in the case of mathematics, these are mostly negative and generally related to certain teachers or to particular topics. It is common for people to admit that they did not enjoy much (if any) of their school mathematics education and, perhaps more surprisingly, for even highly skilled, well-educated professionals to claim that they "could not do it" (see e.g., Cockcroft, 1982) — this was usually around the level that they were at when they ceased its formal study. At different times and in different places around the globe, adults have experienced systems of education driven by different political and philosophical positions.

Over the last one hundred years or so there have been several major schools of thought on education, including adult learning and development, and these are necessarily generalised across formal schooling and various forms of adult education. The categorisations found below are for analytic purposes only and are not necessarily discrete. It appears that as one fades from prominence another replaces it, in a cyclical fashion. In this book I will address five, drawing in particular on the work of Cranton (1992) and Tennyson and Schott (1997) in relation to adult education and instructional design in general, and Paul Ernest (1991) with respect to philosophies of mathematics. I will also reflect upon my own experience of over 20 years teaching mathematics and numeracy to adults in basic, further, and vocational education settings, including workplace-based. I will discuss both general features and those more specific to mathematics; I will also reflect upon the consequences of those paradigms that I have experienced personally.

*Chapter 2*

# FIVE PARADIGMS OF EDUCATION

## 1. LIBERAL/CONSERVATIVE, INTELLECTUAL, PATERNALISTIC TRADITION [OLD HUMANIST]

In general the features include:

(a) a focus on the individual as a psychological being;
(b) unquestioned (i.e. 'politically neutral') cultural transmission of disciplinary knowledge;
(c) the teacher as a figure of authority in the discipline and in full control of the teaching-learning process; and
(d) the delivery taking the form of a lecture, with minimal direct interaction with or between students.

In the case of mathematics:

(a) the discipline is treated as a fixed body of knowledge (i.e., with underlying absolutist/Platonist philosophies);
(b) the pedagogy is transmission-based, with lectures on theory and some worked examples followed by exercises set for the learners;
(c) the aim of the teacher is to explain clearly and to motivate the learners;
(d) the role of the learner is to understand the material and to apply it as appropriate;
(e) between them, the text/s and the teacher are the sources of authority and assessors of correctness;

(f) mathematical subjects or topics are hierarchically organised, and mapped out in advance;
(g) examinations are often externally set for the school or even the education district or directorate; and
(h) learners vary by their innate ability.

This tradition reflects how I was taught at school and then university in the 1950s and 1960s, and is still common in many parts of the world today. The sources of motivation for the adult learner might include the need or desire to become an 'educated person', to gain a particular qualification or entry to work or further study, and/or to learn the language of the discipline.

As an adult educator this was how I expected to teach adults in my first vocational and further education teaching position. All I had to do was to speak nicely to the learners, young and old, and be infinitely patient in explaining and re-explaining. However, it is now apparent that this was just repeating the same processes that learners had probably experienced in the past, possibly in those earlier times with less sympathetic teachers and more pupils demanding the teacher's attention. The focus nevertheless remained on individual responsibility for learning, thus continuing the potential for learner frustration and humiliation. On the other hand, this method could be very suitable for learners who already had a love of learning and a thirst for knowledge. They would need to be already highly motivated, and even willing to seek out further information for themselves — nowadays the internet offers great potential for information seekers but critical searching skills need to be developed. It could also be part of the process of adults proving to themselves that they really 'can do it' now, at this later stage of life.

## 2. BEHAVIOURIST [TECHNICIST] APPROACH TO TEACHING AND LEARNING

Also known a *mastery learning*, in general the features include:

(a) a focus on the individual as a psychological being, with assumptions of fixed ability realised by hard work;
(b) an unquestioned (i.e. supposedly 'politically neutral') cultural transmission of disciplinary knowledge;

(c) the pedagogy is based on so-called 'scientific principles' [Taylorism] and derived from earlier psychological work done by Skinner and even Pavlov;
(d) the curricular content is pre-determined elsewhere, following explicit hierarchies of knowledges and skills, and the content is atomised into minute competencies;
(e) the learner is programmed to move through these individual competencies in a fixed order;
(f) the role of the teacher is to assist students who become stuck and to keep extensive records of achievement; these may be automated in the case of *Computer-Managed Learning*; and
(g) assessment is formative in the mastery learning system, but summative examinations may also be externally set.

This paradigm is associated with a belief among some policy makers and senior bureaucrats that it is possible to devise the 'one best way' of teaching, and there have even been attempts over the years to make materials 'teacher-proof'.

In the case of mathematics:

(a) the role of the learner is to work hard, make an effort, practise continually, and even submit to rote learning where necessary;
(b) the role of the teacher (or surrogate worksheet — all too ubiquitous in adult numeracy education!) is to drill-and-fill the learners with facts and algorithms;
(c) the focus is still on the mathematics and applications are tailored or constructed around the mathematical skills, with transfer assumed unproblematically;
(d) this low level transmission of skills with possible 'applications' is easily transposed into electronic forms of delivery and commonly available online and in CD-ROM versions;
(e) an extreme view would see no room for calculators in the classroom;
(f) testing of the basic facts is decontextualised — if applications are given they are pseudo-contextualised;
(g) the students have often assumed to be monocultural and even gender-neutral (i.e., white and male).

In addition to those of the traditional model, the sources of motivation for the adult learner might include the need or desire to succeed in entry level tests (often timed) such as for nursing, the police, or the armed forces. Students are often motivated by experiencing repeated success from small steps and moving through the sheets as quickly as possible. They may also enrol in such classes in response to political pressures where education is seen as the solution to social problems such as unemployment and where the education system can be highly regulated and held accountable via data such as rates of completion of pre-specified competencies (FitzSimons, 2002).

Competency-based training (CBT), which appeared in Australia in the 1990s and was based on observed performance of workplace competencies rather than underlying knowledge, is strongly linked to behaviourist approaches. In adult mathematics classrooms learners work through individual worksheets and complete mastery tests in order to achieve 'competent' outcomes within the limited time available. Prior to the 1990s it was common for teachers such as myself to work at an individualised level using an assortment of textbooks, some of which even had appealing titles like "Everyday Mathematics" or "Real Life Mathematics". However, this was just repeating the same focus on individual effort — albeit in a different format — which could ultimately become very boring in its social isolation for the learners. This focus on individual responsibility for learning continues the potential for learner frustration and humiliation. Having the teacher in full control removes much of the agency from the adult learner in decisions about content and sequence. On the positive side, it should also be recognised that mastery learning can give a wonderful feeling of achievement — maybe for the first time for many adult learners of mathematics. It can really help to develop a feeling of self-confidence; and the certification is cause for pride in many adults.

Under the first two paradigms, mathematics education epitomises authoritarianism in the guise of simplistic right/wrong decisions — whether in the final answer or even in the necessity of so-called correct workings being shown. In its desire for certainty in an uncertain world, the 'back-to-basics movement' — which would support such an authoritarian approach — is a prime example of the concept of using mathematics/numeracy as a means of social training in obedience under the guise of a need for teacher accountability. Although there are certain justified social and economic needs for accuracy and efficiency throughout history (e.g., in emergency services), these are not universal and emphasising these may still be a source of mathematics anxiety or mathematics avoidance for many adults today.

## 3. PROGRESSIVE EDUCATION OR STUDENT-CENTRED APPROACHES

Stemming from the work done by Dewey in the early part of the 20$^{th}$ century, these approaches reached a crescendo in the 1970s free-schooling movements. These were widely adopted as policy by schools prior to neoliberal or economic rationalist governments in Australia and the UK, and are reflected in much of the international literature on lifelong learning since the 1970s as they assume a substantial degree of learner autonomy.

In general the features include:

(a) a focus on the individual, but within a social context,
(b) an emphasis on reflection and action, and
(c) a curriculum ideally focused on the immediate problems and needs of the learners.

Learning is seen as personal growth for the individual. The teaching methods could include problem solving, scientific method or experimentation, and learning contracts. The facilitator is responsible for minimising the barriers to learning by organising caring and supportive work groups, for example. In other words, the role of the teacher is to prevent failure and to facilitate personal exploration. Learners are acknowledged to have varying abilities but these need to be cherished.

With respect to mathematics, in the 1980s the use of problem solving and small group work was highly fashionable, in English-speaking countries at least, when process superseded content as the main emphasis. There were efforts to have learners work 'as mathematicians' and this eventually led to the adoption of constructivism in mathematics education, where it was acknowledged that learners construct their own knowledge rather than receive it. Much research came out of the USA on how even young children could form communities of learners and negotiate the correctness of mathematics practices. The ideal was to solve, and perhaps even pose, problems in ways that reflected professional mathematics practice rather than the century-old factory model of schooling. Many of the constructivist ideas are still current in documents published by the National Council of Teachers of Mathematics (e.g., NCTM, 2000) and the American Mathematical Association of Two-Year Colleges [AMATYC] (e.g., Cohen, 1995), for example.

The sources of motivation for the adult learner might include the desire to:

(a) develop more fully as person in areas to which they were previously denied access,
(b) learn the ways of thinking of the discipline of mathematics — as mathematicians,
(c) learn that doing mathematics can be fun and a social activity.

Inspiration may emanate from the self, significant others, and perhaps the study-group itself. (See FitzSimons, 2002, chapter 3, for comparison of Bernstein's performance and competence models which reflect the traditional and the progressive models for adult learners discussed so far from a different theoretical standpoint.)

In 1986 I was part of the "Teaching Mathematics to Women" project whose goal was to provide a teachers' manual which incorporated women-friendly examples and activities, encouraging discussion and active participation among all participants. Due to this and to a concurrent Australian schools-based program that recognised and documented the expertise of practising teachers in developing innovative mathematics teaching strategies, I began to have confidence to move beyond textbooks and worksheets in order to work with my students to develop meaningful activities that were of interest to the class.

Prior to the introduction of CBT, in Australian adult education it was permissible and even encouraged to negotiate the curriculum. Mathematics classes were intended to be friendly and welcoming as the students sought to overcome previous alienation or to continue studies interrupted for a whole range of social, cultural, and economic reasons — for some, by war in their former countries. I was very enthusiastic about the constructivist movement and the shift to viewing mathematics as a fallible discipline created by people over time to meet their various needs (Ernest, 1991). The history of mathematics played an increasingly important role in my classes, as did games and experiments with concrete mathematical and everyday objects, graphic computer software programs, and research projects on female mathematicians and everyday news items of interest. Having access to sophisticated problem solving and modelling teacher support materials from Australia and the Shell Centre in the UK, mathematics could be taught in meaningful contexts and there was room for a mathematical way of viewing the world, including its creative, aesthetic, and multicultural aspects. In addition to this, there was space for reflection by the learners on their experiences of past teaching methods and some of their unintended consequences as the students prepared personal mathematics learning histories and focused on the affective as well as the cognitive domain. Discussion of current teaching methods helped those with children at home to work together with them for understanding; mothers could act as role models of females confident with mathematics.

However, in order to deviate so markedly from learners' expectations — even though the traditional format may have been less than satisfactory in their earlier education — I learned that it is important to take the learners with me on the journey of exploration. Learners can indeed become quite angry at not receiving the kind of teaching they expect from their previous education experience: "Just tell me how to do it, and tell me the answer so that I know I am right." (These kinds of expression are still commonly heard by adult mathematics and numeracy educators.) It is necessary to explain what is intended and the justification for doing so — especially when it looks like the class is having fun! The importance of discussion and dialogue cannot be understated. This kind of teaching helped the students in my classes develop a massive boost in self-confidence and a visible sense of agency. One documented example is of a woman I called Marja (FitzSimons, 2003), who migrated to Australia with her family from Holland as a child. She overcame severe shyness and a lack of self-confidence to act as a role model and even tutor for her young children and their friends.

## 4. INDIVIDUAL SELF-ACTUALISATION, SELF-DIRECTION, SELF-FULFILMENT

This movement holds that content is secondary to process and, in the extreme, the adult learner is to be responsible for their own planning and assessment. The teaching methods are experiential, with a focus on discussion. They were informed by humanists such as Carl Rogers (1969) in his work on self-actualisation, reflected in recent trends towards empowerment. David Kolb (1984) proposed four stages in learning: (a) concrete experience, or being involved in a new experience; (b) reflective observation — observing others in an experience, or developing observations about our own experience; (c) abstract conceptualization — creating concepts and theories to explain our observations; and, (d) active experimentation — using the theories to solve problems and make decisions. Clearly, these ideas can still be recognised in current adult education practice. Kolb also constructed a *learning style inventory* to classify learners according to where they seem most comfortable in the learning process.

In terms of *Andragogy*, Malcolm Knowles (1980) established four basic underlying *assumptions* — not rules:

1. Adults are self-directing.
2. Educators need to draw on learners' experience.
3. The readiness to learn depends on need.
4. Learning should be problem-centred.

Cranton (1992) notes that in his later work Knowles stresses that self-directedness is a *goal* of adult education, not a characteristic of adults per se, and that they have a need to become self-directed learners, assisted by adult educators, as they work towards this ideal. However, this process is not always straightforward and learners may progress through a variety of reactions, both positive and negative, and these need to be recognised and addressed by educators at both emotional and cognitive levels. From this perspective of self-actualisation, academic content is secondary. However, the principles may be incorporated into other paradigms.

## 5. SOCIAL TRANSFORMATION

In general the features include:

a   a focus on the individual, but within a social context,
b   education is used to achieve a new social order,
c   the focus is on the collective, with the teacher and learners as equal participants in a group, learning from each other,
d   problem posing and dialogue play an important role, and
e   ability is seen as a cultural product and not fixed.

Emancipatory educators such as Paolo Freire (1972) attempt to produce individual change as well as social change, and see teachers as co-learners to enable liberation from oppression. They stress the importance of dialogue and critical reflection. Jack Mezirow (1996) discusses transformation theory as a reconstructive theory, explaining the structure, dimensions and dynamics of the learning process. It does not offer a cultural critique per se, but offers a model for people to understand how adults learn in various cultural settings. Adult learners may reassess the values and assumptions formed previously and be motivated to:

- develop more fully as person
- make a difference on the local and/or global scene
- gain respect for one's past history and experience.

Cranton (2006) follows Mezirow's perspective but no longer sees transformative learning as an entirely cognitive, rational process. She includes the role of imagination and spirituality in transformation, and the importance of affect in the process. Drawing on the idea of connected knowing and contributions from critical theory, she offers an holistic model of transformative learning. She has also revised her earlier oppostion to electronic delivery of teaching programmes in light of the possibility for improved communications between participants.

From a mathematics perspective, with a democratic socialist philosophy and a social constructivist view of mathematics there is an emphasis on social justice and citizenship. Learning takes place by questioning, decision making and negotiation. Teaching encourages discussion, conflict, questioning of content and pedagogy. Resources are socially relevant and authentic, assessment takes a variety of modes and incorporates social issues and content. The accommodation of social and cultural diversity is a necessity. Critique of mathematics and the role it plays in society are important (e.g., Skovsmose, 1994). Dialogic learning, informed by the work of Raymond Flecha, has been used effectively with parents as adult learners of mathematics in communities in Tucson, USA, by Marta Civil (Civil & Quintos, 2009) and in Barcelona by Javi Díez-Palomar (2007). (See also the Special Edition of *Adults Learning Mathematics – An Online Journal*, *3*(2), November, 2008. Available at: www.alm-online.net.)

In FitzSimons (2000, 2001) I discuss a hybrid model I devised when at a pharmaceutical manufacturing company where workplace teaching was undertaken within constraints of CBT learning outcomes, company work priorities, and ethical considerations. (This will be discussed further below.) Here, the motivations of individual workers included the need to:

- gain a credential in recognition of the work they were already competently performing
- keep their current job (under subtle but persuasive pressure from management)
- possibly gain a promotion or the opportunity to switch jobs
- prove they could 'do it'.

The motivation of management was to:

- increase productivity
- comply with the German headquarters ethic of educating all workers
- assist the workers to become familiar with the vocational and higher education system in Australia
- possibly find workers worthy of higher duties, e.g., in-house training roles.

One positive outcome from my point of view was that the manager observed an increase in worker confidence and participation in workplace discourse. Another was that, once they had overcome their initial fears and resistances, the workers were proud of their achievements.

This section has outlined five major paradigms, based on different philosophies of education, which have been drawn upon to inform policy and practice in formal education settings for lifelong learning, as well as the school sector. As noted above, different kinds of formative experience will have different consequences for adults as they continue on their unique pathways of lifelong learning. Although the paradigms have been addressed discretely, it is possible for adult educators to draw on two or more in any given teaching programme in order to meet the needs of their particular students within the constraints afforded by the programme and the conditions under which they are engaged. I now turn to focus on the distinctions between mathematics and numeracy because the terms are often used as synonymously. I will argue that mathematics alone in necessary but not sufficient for lifelong learning as described in the Introduction.

*Chapter 3*

# DISTINCTIONS BETWEEN MATHEMATICS AND NUMERACY

In some policy documents and the popular media it is common to see the terms mathematics and numeracy used interchangeably — as if the latter was a more 'user-friendly' version of the former. There is an implication that mathematics is serious and has high social value while numeracy is trivial, for those less able, and is really 'just common sense'.

In FitzSimons (2002) I examined the literature on various public perceptions of mathematics, including those of employers. It is generally considered that a certain level of qualification is important, even if only for the purpose of gatekeeping to discriminate against those individuals who lack such credentials on the grounds that they act as a proxy for some kind of intelligence. It is common that, when asked to document the mathematical skills appropriate to the vocation or occupation of interest, employers typically regurgitate lists of topics that they remember from their own school days. It is common, also, for people outside of the mathematics education world as well as within, to hold an absolutist philosophy — generated and reinforced over many years by traditional school mathematics education — that mathematics is objective and value-free. In recent decades at least, there are many professionals within the field, especially researchers, who hold that mathematics is neither objective nor value-free and, moreover, that it is a fallibilist discipline — as can be seen from the many and continuing changes to its truths over time. It seems reasonable to assume that most of the population in the so-called developed world still believes that in mathematics there is one and only one correct answer and, most likely, one best method of finding it no matter what the circumstances. This belief is likely to be accompanied by expectations that the traditional, transmission model of teaching is the most appropriate, even though it has failed so many people over the last century or so of mass public education. On the other hand, more sophisticated understandings of the mathematics used in the workplace and everyday life have followed from recent research, often based on activity theoretical frameworks where the complexities of the processes and interrelationships are illuminated (see, e.g., Bakker et al., 2006; Hoyles et al., 2002).

Following the theoretical work of Basil Bernstein (2000), in FitzSimons (2008) I argued for the distinction between mathematics and numeracy. Numeracy, whether for adults or children, is a social construct, drawing upon mathematical skills and knowledges developed over a lifetime. These may be learned informally from — and even taught by — family, friends, and other significant others. In the case of pre-school children and unschooled youth and adults, learning is dependent on the social and cultural settings available to the learner in their various communities of practice. In countries where formal education is the norm, funded to a greater or lesser extent from the public purse, decisions are made by governments (advised by bureaucrats) about the quantity and quality of education for various groups of learners — the voices of certain groups of stakeholders are privileged over others, depending on the political complexion of the government of the time. The major focus may be on, for example, improving business or national economic outcomes or on democratic citizenship, or some combination of these. Whatever the focus, numeracy is necessarily related to mathematical skills and knowledges.

Bernstein (2000) describes the discipline of mathematics as being a *vertical discourse* due to its coherent, explicit, and systematically principled structure. There are many sub-disciplines (e.g., algebra, geometry, trigonometry), each with their own specialised and codified languages, and these are generally taught, in formal school education at least, as separate entities. Starting with simple ideas, layers of complexity are gradually added over the course of an education. However, sometimes apparently simple ideas such as "two negatives always make a positive" can be misleading when (mis)remembered outside of the context of application (i.e., it applies to multiplication and division of signed numbers, but not to addition and subtraction). In formal education, the discipline of mathematics is recontextualised for the purpose of enculturation. Just as the school subject of woodwork is qualitatively different from the trade of carpentry, so school or formal adult mathematics education is different from the practices of professional mathematicians or statisticians. It is also quite different from workplace numeracy, based upon my own practical research into how mathematics is used in the workplace as well as literature reviews (e.g., FitzSimons, 2002; FitzSimons, 2005, FitzSimons & Wedege, 2007). In school, the object is to learn mathematics and, hopefully, to gain a suitable qualification; in the workplace it is to get the job done.

Following Bernstein (2000), I argue that the construct of *numeracy* is an example of a *horizontal discourse*. Bernstein describes these as being a set of strategies which are local rather than global (as with modern day mathematics), compartmentalised (e.g., the form of counting applied in one situation may not be relevant or transfer to a different situation or set of circumstances), context-specific and -dependent, aimed at maximising encounters with people and the environment. Such knowledges are embedded in on-going practices, usually matter to the user on a personal level and are directed towards specific, immediate goals, highly relevant to the user in their own life context. This definition bears a strong affinity with research reports on workplace numeracy and everyday activities involving the use and re/construction of mathematical knowledges (FitzSimons, 2008).

Following Bernstein (2000) once more, compared to the discipline of mathematics, numeracy is weakly classified in terms of its necessary integration with context. Whereas there is a sharp division between mathematics and most other school subjects and life outside the classroom, numeracy literally depends on making links with the non-mathematical specific context. The aim of formal mathematics education is for the learner to be able to form mathematical generalisations and abstractions that override individual experiences and examples, and to make what might be regarded as rational scientific judgements or predictions accordingly. On the other hand, numeracy depends specifically on an individual's experiences and their knowledge of the particular context at hand in the formation of their judgement. Decisions may include reference to various sensory inputs as well as expectations based on knowledge of relevant historical data. Whereas in mathematics there is a well-known hierarchy from common sense up to so-called uncommon sense, in numeracy common sense is of the essence. High level abstractions alone are insufficient and may even prove counter-productive when they cannot be related to the actual problem at hand. Numeracy cannot be said to have a specialised language, except at the most local level of use in context. For example, the use of the term "thou" [i.e., thousandths] is widely used in the building and automotive industries, but may not have meaning elsewhere. Numeracy is not necessarily explicit or precise (but can be if required), and its capacity for generating formal models may be limited to the context at hand rather than more widely generalisable.

In essence, then, numeracy is a *horizontal discourse* which draws upon foundations of mathematical knowledge developed by individuals over a lifetime of personal experience and enculturation but which, unlike the *vertical discourse* of the discipline of mathematics, relies on common sense and is context-specific and -dependent, directed towards the achievement of specific, immediate, and highly relevant goals. It is even possible for a person with high-level qualifications in mathematics not to be numerate in certain spheres of everyday life or other contexts with which they are unfamiliar. In practice, the most sophisticated mathematical solution may not be the most useful, or even feasible at all.

## PEDAGOGICAL IMPLICATIONS

Vertical discourses such as mathematics consist of specialised symbolic structures of explicit knowledge; procedures are linked hierarchically. That is, they start with the simple and the concrete (where possible) and progress towards abstract generalisations. The formal pedagogy is directed towards some unspecified projected application (e.g., "you might need this next year, or at work") and is an on-going process, generally continuing over an extended period of time (usually the compulsory years of schooling at least).

By contrast, according to Bernstein (2000), the pedagogy of horizontal discourses is usually carried out through personal relations, with a strong components related to emotions and beliefs. It may be tacitly transmitted by modelling or showing, or by explicit, didactic means. The pedagogy may be completed in the context of its enactment, or else it is repeated until the particular competence is acquired. From an individual's perspective, "there is not necessarily one and only one correct strategy relevant to a particular context" (p. 160). (This is stark contrast to the traditional teaching of the "one best method" in mathematics.) Bernstein concludes that horizontal discourse "facilitates the development of a repertoire of strategies of operational 'knowledges' activated in contexts whose reading is unproblematic" (p. 160). In the case of numeracy, the person has a range of mathematical (and other) relevant options and decides which might be the most appropriate for the particular situation at hand.

In summary, whereas the transmission of formal mathematics knowledge is likely to progress from the concrete to the mastery of simple operations, to more abstract general principles, the teaching of numeracy to adults may have more in common with the reverse processes which take place in workplace learning. In other words, general principles are understood but need to be made concrete in order to be realised. (See the following section on Activity Theory for further discussion.) Ultimately, the learner will be expected to develop a repertoire of context-dependent strategies, based on experiential learning from a more 'knowledgeable' person (or persons) in a given situation, where achieving the task itself is the priority — not the learning of mathematics per se. Context-specific and localised models of action may also be developed and practical knowledge/expertise, together with common sense, is highly valued.

The construct of numeracy has only been in use, in English speaking countries at least, since the 1950s. In different countries different terms are used to apply to the use and even construction of mathematics that has direct relevance and application outside of the classroom. *Numeracy* is often used with the pejorative tag *basic* attached in the case of adult education and even today this often conjures up expectations of early years school work albeit with 'adult' examples inserted where possible. When the term numeracy is applied generally to education in campaigns, it may be that this term is being used in an attempt to popularise it with school children and their parents by disassociating it from the cold, hard, judgemental image of the discipline of mathematics. Other terms in use are *quantitative literacy*, *mathematical literacy*, and also *functional mathematics* (see FitzSimons, 2008a). However they are labelled, the curricular content, pedagogy, and assessment still remain political phenomena. The intended outcome of mathematics and numeracy teaching is *numerate behaviour* but, as I have argued above, a sole focus on the vertical discourse of mathematics alone will not guarantee such behaviour.

My own experience 20 years of teaching vocational mathematics and statistics in mainstream Australian vocational education institutions [originally known as Technical and Further Education or TAFE] (see FitzSimons, 2002) showed the dominance of statewide curricula, then national learning outcomes under CBT. Aspects of the discipline of mathematics were selected somewhat arbitrarily (often based on employer's opinions rather than scientific research) and turned into manageable chunks of theory accompanied by (pseudo)contextualisations of varying degrees of relevance and meaning to learners. Some of the contextualisations were pure fantasy, others required procedures that had not been used in the relevant vocation for years, certainly since the introduction of modern calculators and software packages. This theory was generally followed by sets of exercises of comprising decontextualised skills and, eventually, so-called applications where the intended concepts were dressed up in word problems intended to ensure 'transfer' at some future point in the learners' lives.

In spite of my best efforts at helping learners to make meaning by adopting various pedagogical approaches including practical activities with concrete mathematics and everyday resources, videos, and industrial and educational computer software packages, I had limited success. Some students continued to make serious errors in their practical classes, as they made no attempt to reconcile ridiculous answers derived from mistaken calculator-generated 'answers' with their experience of practical classes in the laboratory. The situation for learners who lack ongoing personal contact with teachers and possibly other learners — as in the case of technology-delivered courses (e.g., via online or CD-ROMs) — can only become more challenging (see OECD, 2005, chapter 4). On the other hand, numeracy learned from and through the workplace or everyday life experience necessarily takes place in social contexts where others are present, surrounded by mediating artefacts such as tools, equipment, machinery, manuals, documents for instruction and reporting (FitzSimons & Wedege, 2007). There are also computer packages for purposes of communication, for data collection, analysis, interpretation, and presentation, and for location of people and objects through databases. This is in addition to internet mediated information searching and retrieval. In other words, as citizens and workers, learners are immersed in social, cultural, and historical contexts of meaning, and serious mistakes in the realm of mathematical calculations or judgements are comparatively rare (see, e.g., Lave's 1988 discussion of shoppers, weight-watchers, etc.).

In this section, the theoretical work of Bernstein (2000) has been used to differentiate between mathematics and numeracy, and this relates to the object of teaching and learning activities for lifelong learners. In the previous section, I outlined and discussed five major education paradigms that have featured in adult education, some of which adult learners are likely to have experienced in their years of compulsory education. As noted previously, how adults have been taught in the past is likely to shape their attitudes towards and expectations of further learning, especially in formal education settings.

Because of the importance of adults' education and work/life histories as well as their socio-cultural contextual settings, it is useful to seek a theoretical framework that can incorporate these significant aspects of the lifelong learning process. In the following chapter I will give a brief history and an overview of social, cultural and historical activity theory. After this, I will draw on the work of Yrjö Engeström as well as Kari Kuutti to draw out salient aspects for mathematics (or numeracy) learning in general, with implications for lifelong learning in a technological era.

*Chapter 4*

# PERSPECTIVES ON ACTIVITY THEORY

Hedegaard, Chaiklin, and Jensen (1999) reflect on the work of the Soviet cognitive psychologists, Vygotsky and Leont'ev among others, whose work was published around 1930. Human development can be seen as a social and cultural-historical process due to the fact that humans pass on tools and procedures for their use (i.e., knowledge) to the next generation. Vygotsky's genetic law of cultural development states that every function appears in two planes: firstly on the social plane as an intermental category, and secondly, on the psychological plane as an intramental category. This is elaborated by Daly and Mjelde (2001) who hold the assumption that meaning arises from social interaction arising from outside of the person who takes it in, then sifts it and transforms it by accepting, modifying or rejecting it, according to personal experience and common cultural values, and finally responds.

Hedegaard, Chaiklin, and Jensen (1999) continue that:

> Leontiev conceptualized activity as a collective process, with actions as goal-oriented processes of individual subjects, and operations as psychic functions conditioned by the prevailing material conditions and available tools. ... Work was taken as the prototype of activity, and other types of activity were developed through human history as derived from work. (p. 14)

For them, social practice lies at the heart of the theory's conceptual structure, even though they accept that the concept is not unique to the cultural-historical tradition. They believe that the activity concept (and associated concepts such as action, motive, goal, leading activity, and motive hierarchy) make it possible to provide an elaborated set of concepts that can be used to give a differentiated theoretical analysis of social practice.

Stetsenko (1999) asserts that the concepts of social interaction, cultural tools, and the zone of proximal development [ZPD] — where a learner, in collaboration with a more experienced peer, is able to solve problems on a more advanced level than when acting alone — are all integral to a person's development. She notes that cultural tools and the "complex systems that are formed by them, include different kinds of numbering and counting, mnemotechnical aids, algebraic symbols, art works, writing, schemes, diagrams, maps, drawings, and all sorts of signs" (p. 236). Stetsenko stresses the need for the synthesis of these three concepts — that is, the acquisition of cultural tools through social interaction in the ZPD —, together with their respective research traditions, as an important and continuing task in activity theory. From this perspective, then, learning then plays a leading role in human development, as its essence, rather than merely playing a supportive role.

Drawing on the long tradition on Soviet cognitive psychology in the 20$^{th}$ century, Engeström (1987) formulated what he termed the basic mediational triangle (Figure 1) as a model to support the description and analysis of learning from a social, cultural, and historical activity theory perspective. Learning is seen as a collective process, dependent on interaction and communication. Engeström holds that the concept of learning activity can only be constructed through a historical analysis of the inner contradictions of the currently dominant forms of societally organised human learning. Tensions and contradictions play a major role in the learning process.

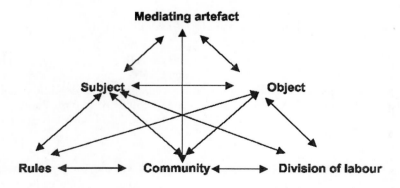

Figure 1. The basic mediational triangle expanded (after Engeström, 1987).

The upper sub-triangle represents psychological traditions which focus on the learner or group of learners (subject), the intended goal of learning (object), and the things used in the transformation process, including both material tools and tools for thinking (mediating artefacts). The remaining sub-triangles take into account the social, cultural, and historical factors, namely the explicit or implicit rules or norms that regulate the processes of behaviour and, in this case, the discipline of mathematics, the people who are involved in the process (community), and the explicit or implicit organisation of that learning community — or who does what — (division of labour). Engeström (1987) notes that these vertices form part of a dynamic system, with each vertex in a continual state of contradiction or tension with all others — indicated by the double arrows. Finally, the learning outcome is the net result of all of interactions in the activity system. This diagram may be utilised to describe and analyse a mathematics or numeracy class activity for adults whose motives are to seek employment or progress to further study. In a similar fashion it may also be utilised by a group of teachers intending to become adult numeracy specialists. Or it may be used to describe and analyse the teacher's perspective within their institutional constraints, and so forth. The specific learning object might be, for example, to understand the structure of the metric system in relation to the decimal number system and to use it for purposes of calculation in contexts which are of interest to participants.

According the Engeström (1987), the essence of learning activity is production of objectively, societally new activity structures (including new objects, instruments, etc.) out of actions manifesting the inner contradictions of the preceding form of the activity in question. Learning activity is mastery of expansion from actions to a new activity. In other words, Engeström sees the endpoint of learning not as the reproduction of facts and skills, but that these might contribute to a new activity as the particular learning community struggles to overcome the tensions and contradictions of an existing but problematic activity.

Engeström (1987) continues that the object of learning activity is societal productive practice, or the social life-world, in its full diversity and complexity. Productive practice, or the central activity, exists in its presently dominant form as well as in its historically more advanced and earlier, already surpassed forms. An example could be found in the contemporary schooling for adults (and children) which takes place in historical contexts of buildings, with mediating artefacts such as resources and discourses, and so forth, founded in former social eras but with modern-day pressures and expectations placed upon teachers and students who must reconcile these tensions and contradictions and find new ways to achieve the desired outcomes. The introduction of electronic technologies into mathematics (and other) classrooms has exacerbated these tensions and contradictions.

Following Engeström (1987), in the workplace or formal educational setting, at first the learner sees the object in the form of discrete tasks, problems and actions. The learner must then discover the goals of the activity — and this can be problematic for those whose backgrounds lie outside of the mainstream learning culture in that particular setting — for example, new workers, newly arrived immigrants, and diverse social and cultural sub-groups. The learner must analyse and connect these discrete elements with their systemic activity context, and then transform them into contradictions demanding a creative solution — that is, to see the problem that needs resolution. They must then expand and generalise their actions into a qualitatively new activity structure within societal productive practice — that is, devise a new method of working in this situation (i.e., until new tensions and contradictions arise). Engeström notes that these components result in a theoretical reconstruction of the object.

In the case of adult and vocational mathematics/numeracy education, the central components are the learner as subject, mathematical or numerate activity as the object, with technology and other tools (mathematical and otherwise) as the mediating artefacts. Technology also supports a range of mathematical tools for exploring new possibilities, developing understanding, maintaining skills, communicating with others, and acting as a labour-saving device. These are all set in the socio-cultural context of rules, community, and division of labour. Engeström (1987) proposes that there are multiple levels of contradictions so that each vertex may be expanded to form a new activity system with its own tensions and contradictions. He underlines the fact that the activity systems of learners and teachers necessarily have different objects — a point which is often overlooked in the pedagogical design process. Engeström also observed the essential peculiarity of the formal education activity of students in its strange 'reversal' of object and instrument. In wider societal practice, text (including the text of arithmetic algorithms) appears as a general secondary *instrument*, whereas in traditional school and vocational mathematics education settings the usual *object* of learning is manifested in the production of texts of various modalities (i.e., numerals, algebra, graphs, charts, tables, statistical symbols) rather than the production of concrete or tangible artefacts — as in service industries or trades, for example.

In relation to the introductory discussion of lifelong learning, Engeström (1987) made several further observations about learning, founded on his very strong belief that the motivation must emanate from the individual and their personal interests and goals, rather than being imposed and structured from outside: Learning must be meaningful for the learner. Accordingly, he is critical of much of the commonly accepted rhetoric on teaching and learning. Firstly, he is critical of the teaching of problem solving in standard reactive forms, which present the individual with a pre-set learning task. He claims that learning is defined so as to exclude the possibility of individuals finding new contexts, causing difficulties in anticipating, mastering and steering qualitative changes in their lives, in families and organizations, and in society as a whole. While so-called poor learners are being helped to cope with tasks given to them, at the same time the contexts of those tasks and skills are undergoing qualitative changes that render previous ones obsolete. This is particularly obvious in the workplace where change and the search for improvement are critical features (FitzSimons, 2008b). Secondly, Engeström is also critical of traditional methods of teaching which stress the act of memorising thereby assuming a constancy of knowledge, and implying the non-developability of a person. Thirdly, he notes that traditional logic, such as concepts and propositions (of the kind found in mathematical discourse) which must remain rigid each time they are repeated, lies in stark contrast to the live thought processes which are engaged in order to understand and to improve a problematic situation. Engeström observes there is a general shift of skills away from routine treatment towards diagnosis. In other words, creative thinking is required to solve the ever-evolving problems in life and work situations which are characterised by uncertainty, disorder, and indeterminacy.

Discussing metacognitive skills, Engeström (1987) argues that they do not exist in a vacuum and is critical of those who recommend reality testing ("does this make sense?") when it is quite possible that the task itself does not make sense to the learner. Learners in regular classrooms, especially when they are being purposefully instructed in problem solving or metacognition, are not often given the opportunity to create new situations for themselves. He argues that to develop a useful level of metacognitive awareness requires conscious analysis and mastery of discrete learning situations, as in the classroom, but also of a continuous activity context where the situations are embedded, whether in school or out, in the workplace, and so forth.

Critical of the scaffolding metaphor as spatial rather than temporal as part of a living process, Engeström claims that it is restricted to the acquisition of the given. By contrast, the concept of zone of proximal development (ZPD) means that teaching and learning are directed towards developing historically new forms of activity, beyond the acquisition of existing or societally dominant forms. This implies following the learners' interests outside of the classroom, and also developing expansive learning activity (i.e., questioning and dialogue) in and between the learners. He reminds us that the instructor's task and the learner's perceived task are seldom identical.

The logical consequence of Engeström's work is that knowledge is not simply 'transferred' from teacher to learner or from learner to application. Knowledge is constructed, and transformed to meet the needs of the fully contextualised situation at hand. From the perspective of the ultimate participation in workplace and other civic activities, there are compelling arguments for the importance of collaboration in mathematics/numeracy education, particularly where it is, or can be, mediated by information and communication technologies (ICTs). Following a similar line of reasoning, collaborative learning is also an important component: In the learning situation collaboration can go well beyond the realms of formally enrolled class members in the same room to extend to learners at a distance in time and/or location, as well as to include significant others in the community. Finally, the division of labour suggests a rethinking of the roles and responsibilities in the learning process as part of the shift away from the traditional transmission paradigm of teaching and learning.

From consideration of Engeström's version of activity theory and observations on concepts commonly used in formal education I now consider the use of technology in lifelong learning related to mathematics and numeracy. In Figure 1, the vertex labelled *mediating artefacts* encompasses the technologies used in a mathematics or numeracy classroom. These include the traditional technologies of pencils, rulers, protractors, compasses, and so forth. They also include electronic technologies which may be used as labour-saving and sometimes exploratory devices (e.g., calculators), as preparation for the world of work as well as civic life in general (e.g., spreadsheets, word processing), as teaching tools in their own right (e.g., software packages for the development of specific skills), or as a means of communication, locally and globally (e.g., internet searching, email communication). The critical feature of mathematics or numeracy instruction is that technology has these multiple aspects — as one possible medium of delivery and communication, and as an integral tool for developing conceptual understanding, as well as assisting in the thinking process, — and it is this complexity which sets the field apart from many other discipline areas. In particular, there is a need to focus on the use of technology as a tool, but also as an object of learning in its own right. At the very least there are rules governing the use of technology to be learned or invented so that the artefact may become useful and transparent as an instrument (Trouche, 2004). As with the more ancient technologies, each of these needs to be made the focus of learning until its use become unconscious so that it can then support further learning.

*Chapter 5*

# ACTIVITY THEORY AND TECHNOLOGY IN EDUCATION

In this section I will elaborate on how Kuutti's work relates to mathematics education for adults at each of the three levels outlined above. This is important because many well-intentioned mathematics programs for adults, including those utilising ICTs, are not sufficiently theorised so that there is a preponderance of cognitively low-level material which generally neglects contexts of importance to adults as well as the affective domain.

Kuutti (1996) synthesised the work of Leont'ev and Engeström to devise a two-dimensional framework for analysing the use of technology from an activity theoretical perspective. One dimension incorporates Leont'ev's three hierarchical levels of unconscious *operations*, goal-directed *actions*, and collective *activity* with an overarching motive; the second dimension incorporates Engeström's six components of the mediational triangle, discussed above. I have found this framework very useful in my research on adult numeracy and new learning technologies. In particular, it has helped me to analyse the various technology-mediated programs for adult and vocational learners, such as those available as online courses or CD-ROMs. (For further information see the web page for Adult Numeracy and New Learning *Technologies* <http://www.education.monash.edu.au/research/projects/adult-numeracy/>, an Australian Research Council Post-Doctoral Fellowship Research Project [2003-2006].)

## OPERATIONS

At the level of operations, the tools or artefacts are intended to support the automating of routines. The object is to provide data, and to trigger predetermined responses in the learner (or subject). The intention is that sets of rules be embedded and imposed (e.g., for arithmetic/statistical calculations including knowledge and use of the metric system; also algebraic conventions, geometric facts, etc.). An implicit community would be created by linking the work tasks of several people together (e.g., learners on and/or off campus, significant others for the learner). A certain division of labour would be embedded and imposed (i.e., a clear demarcation of tasks between teacher/tutor and learner; also between them and external instructional designers, including regulatory authorities, or programmers). The operations level of support is oriented towards technical mastery.

At this level, the learner would be given feedback in terms of right/wrong answers with some remediation but within a drill and practice format. On the assumption that these operations are already understood by learners, technological and other mediating artefacts such as software programs or hard copy worksheets, could be used to provide drill and practice in number, measurement, algebraic manipulation, geometric facts, calculus, and so forth. In my experience, this is where the majority of commercial resources are located and, while valuable in themselves, should not signify the sole outcome of adults' mathematics education when they return to study. This is particularly the case where the exercises are decontextualised or pseudo-contextualised (FitzSimons, 2002) because they do not support adults in transforming these skills into everyday and workplace practices. This basic level of support may also include reinforcement on how to use calculators and/or computer applications such as spreadsheets or databases, web search engines, or software packages such as dynamic geometry, statistics, symbolic algebra systems, and so forth — once these have been initially mastered. The data could come from text, word, or spreadsheet files, CD-ROMs, or the worldwide web, for example.

There are many programs, technology-mediated as well as paper-based, which operate at the operations level where they are designed to embed and reinforce various forms of mathematical rules, in the typical 'drill-and-practice' format. Of course it is important that certain mathematical facts become automatic in order to enable further development and deeper conceptualisations, but on their own they are not sufficient to support workplace or other social practices and, even though such 'basic skills' are frequently used as testing mechanisms for prospective employees (often as timed tests), they do little to support understanding and communication which are both essential aspects of life.

## ACTIONS

At the action level, there are programs designed to supporting sense-making activities through transformative and manipulative actions related to the concepts intended to be learned. This level also addresses the issue of making the tools and procedures or rules themselves visible and comprehensible. The object of learning itself is also intended to be made manipulable by the learner. Accordingly, supporting sense-making actions within an activity means that materials need to be culturally appropriate, particularly if the program is to be exported — whether online or via CD-ROM or by offshore teaching (FitzSimons, 2007). The rules themselves need to be made visible and comprehensible. In terms of the community, communicative actions need to be supported and the network of actors made visible, particularly if face-to-face meetings are not always possible. This may be via email, discussion groups, online conferencing, and so forth. Concerning the division of labour, the work organisation also needs to be made visible and comprehensible. This means making the various roles and responsibilities clear, which is especially important for adults who may have pre-formed opinions about the way a mathematics/numeracy class ought to be conducted, even if this process resulted in unsatisfactory outcomes for them at some time in the past. Of course, it is also important for learners who — for reasons of geographical location, institutional policy, physical or social disability, incarceration, etc. — work at a distance and may be unable to pick up the contextual clues relating to institutional norms and values.

Face-to-face teaching has been the traditional means of instruction, and teachers of adults have traditionally drawn upon available tools for support, whether they be domestic objects such as kitchen and hardware artefacts; specialised mathematical objects such as number tables, counters, blocks, measuring devices, rulers, protractors, compasses; or hand-held calculators. It is important for teachers to realise that none of these mathematical objects is inherently transparent to learners, even though they are easily taken for granted in contemporary mainstream society. Thus, for some learners, these objects need to be to focus of attention in themselves before they can be utilised for the purpose of enhancing understanding from both cognitive and affective domain perspectives. Consider, for example, the case of newly arrived immigrants from developing countries who bring vastly different social and cultural experiences, or the case of older learners returning to study with no prior experience of the use of electronic technologies in educational settings, and possibly some degree of apprehension and resistance.

There are sophisticated programs in electronic and paper-based forms which address the issue of mathematical meaning. Examples of technological teaching/learning artefacts at this level include dynamic geometry systems, exploratory and interactive educational algebra, statistics, graphing packages — commercial or custom-made. However, as indicated above, it is common to overlook the challenges for adult learners as well as younger learners in becoming confident with using technological tools themselves and understanding their capabilities as well as their limitations. The action level is also intended to support communicative actions and to make the network of actors and work organisation visible. However, these aspects are also frequently overlooked in the focus on conceptual understanding.

## ACTIVITIES

At the level of activity, the focus is on creative, innovative, and transformative learning. This level is designed to enable the automation of a new routine or the construction of a new tool (as, for example, when students make their own teaching aids or authored software programs). The object of learning is enabled to become a common object, shared between learners (as in co-operative learning). The learner is supported in learning and reflection with respect to the whole object and activity (i.e., metacognitive and transformative learning), and the negotiation of new rules is enabled (as is the case for workplace mathematics where local rules may be established). Also enabled are the formation of new communities and the reorganisation of the division of labour (ideally with the learners taking more responsibility for their own learning).

While the process of learning a new routine — which will ideally move down to the unconscious level of operations — is common to mathematics education generally, the other creative and reflective aspects are less so. In the case of mathematics education, the reflection may entail a critique of technology and the uses to which it is put as well as critical mathematics education where the institution of mathematics itself may be the object of critique in all of its social, cultural, and political implications.

Communication, the essence of workplace and other numeracy, may be enhanced by genuine collaboration via projects, local and global, supported where necessary by electronic technologies such as multimedia conferencing (see Borba & Villarreal, 2005, for an example of what is possible as humans work with technology in Brazil, a so-called developing country). The activity level of support is focused on what Engeström (2001) terms *expansive learning* — which allows for creativity and interaction arising from tensions and contradictions within and between activity systems. An example might be where learners realise that they need to know how to work with or modify a technology new to them but have to first focus on mastering it before they can comfortably use it to achieve their prior goal/s.

In summary, the operations level is focussed on automation of routines so that they become unconscious, the action level is content focused and oriented towards developing an understanding of tools and processes, and the activity level supports creative, innovative and transformative learning. So what does this framework mean for lifelong learning in relation to mathematics and numeracy?

*Chapter 6*

# IMPLICATIONS OF AN ACTIVITY THEORY FRAMEWORK FOR LIFELONG LEARNING IN MATHEMATICS AND NUMERACY

In this chapter I will discuss vocational education specifically and also draw implications for adult numeracy in general. In my experience, vocational mathematics education is seen by the learner as more or less useful depending on the quality of the curriculum and the learning activities. Oftentimes it is impossible to qualify as a tradesperson without this compulsory subject — and this can come as a surprise to many students who thought that they had finished with mathematics once they left school. The history of mathematics indicates that it evolved as a useful tool to meet the needs of (sections of) societies over time, overcoming the limitations of human memory in recording information on measuring and allocating wealth, for example. Academic mathematics came into prominence as a result of a confluence of socio-cultural conditions such as the rise of commerce, the invention of the printing press, and the need for labour-saving devices such as counting frames and algorithms. It is only in recent times that mathematics has become recognised as a proxy for intelligence and, with the development of the factory model of mass education, has mathematics education been experienced by many as an instrument of humiliation, either in the short term or over a sustained period. Prior to entering formal education, and certainly outside of it, most people learn and use mathematics naturally and efficiently — in situated and contextualised ways. However, in contemporary society these skills and knowledges need to be further developed to meet ever-evolving technological, economic, cultural, and social demands.

Over the years, well-intentioned mathematicians, vocational mathematics teachers, and employers have inspected workplaces and found examples of mathematics which conform with their traditional views of a 'school-based' curriculum framework — one that is arbitrary in the selection of topics, although the discourse of each individual language (e.g., arithmetic, algebra, geometry) is hierarchical and needs prior skills to become routinised before more advanced ideas can be understood. However, there is no necessity for the imposition of arbitrary levels of what children of given ages (and even adult learners) are expected to achieve across all topic areas in the mathematics curriculum and beyond. My experience of teaching adults returning to study in vocational and non-vocational programmes shows that they can engage with the big ideas of mathematics such as change and approximation, logic, and randomness which might be found at higher levels in school curricula — even though they were nominally located at lower levels in terms of formal assessments of decontextualised mathematics skills. For example, one group plotted the official times of sunrise and sunset (allowing for daylight saving) over a year and this resulted in a trigonometric function. The big ideas of maximum, minimum, and differential rates of change were experienced by them personally in their everyday lives, and so were able to be related to these graphs in a meaningful way. In constructing these graphs, we were enacting the theory of Galperin (cited in Stetsenko, 1999), where

> efficient cultural tools are learning materials (i.e., concepts, theories, ideas) that (a) embody and reify the most efficient cultural practices of the previous generations, in that they (b) express, in a condensed, generalized form, the essence of certain classes of problems or phenomena by (c) representing the genesis of these problems or phenomena and hence the logic — the templates of action and the contexts where these actions can be applied — necessary for dealing with them. (p. 241)

Stetsenko gives the following examples of efficient use of cultural tools. According to this theory, a circle definition is more usefully described as the production of a locus with a stick where one end is fixed or by a pair of compasses, rather than as a curved line with a set of points equi-distant from a fixed point. The first way describes the historic production of a circle, reified in the cultural tool of a pair of compasses. A second example is the concept of a unit, which is more usefully seen as an expression and reification of the operation of measurement rather than as a rote-learned rule. After consideration and discrimination of the various properties of an object, an appropriate measure is chosen. Measurement is then brought into play in order to make comparisons by various properties (e.g., length, weight), and only then the concept of unit (as equal to the chosen measure) and number (as the ratio of some quantity in relation to the given unit of measurement, i.e., *rational* numbers). In ways such as these, learners of all ages can develop embodied and meaningful number and spatial sense in ways that reflect the history of mathematical activities such as counting, measuring, and locating. In fact I have witnessed children in early grades using similar mathematical ideas when introduced to measurement through arbitrary units, such as their foot-length or hand span (cf. non-metric units).

Previously (FitzSimons, 2008b), I have advocated dialogue between vocational trades and mathematics teachers, but this is inadequate where one or both parties hold absolutist philosophies of mathematics and see transfer as the unproblematic 'application' of techniques learned in the activity system of the mathematics classroom. There continues to be massive production of 'Mathematics with Applications' texts, often with CD-ROMs attached, for the vocational and higher education markets. And employers continue to complain that new graduates cannot use the mathematics they supposedly know — according to the novices' qualifications on paper at least.

Instead, there is a need for authentic project-based co-operative work, with fundamental mathematical (and other) processes incorporated throughout the program, and where vocational trades and mathematics (and other) teachers are involved in co-teaching (see, e.g., Borba & Villarreal, 2005, for discussion of integrated mathematics projects in a technology-oriented program). Mathematics learning is then a natural and integral part of the activity, not quickly passed over as a 'tick-box' competency. Rather, adult numeracy and vocational mathematics education need to be seen as a means of enabling adult learners to expand their learning now and in the future, via the path of lifelong learning, as contexts evolve. This proposal will require a different preparation of qualified mathematics and numeracy teachers to be appropriately trained for this different kind of teaching. They will need to have developed a flexible repertoire of mathematics teaching skills combined with adult learning theories, and should be willing to develop a deep understanding of the particular trade or vocation, as well as to work co-operatively with other professionals and industry personnel rather than in isolation. (This resonates with Bernstein's horizontal discourse, utilised in the earlier discussion of numeracy.)

Recently, I have had first-hand personal experience of observing the kind of holistic skills that tradespersons need, as we installed central heating into an older house. It reinforced my previous observations of several workplaces. At every stage of the process were mathematical ideas and techniques, often implicit, but which were critical to the success of the project. Mathematics was involved in all of the other widely accepted generic competencies such as planning and organising, problem solving, working co-operatively, communicating effectively— starting from discussions of the concept, the costing, the agreements on costs, dates, locations and dimensions of heaters, etc. with us, the customer. It also involved the contractor organising his workmen, the correct materials to arrive on time, the logical order of preparation and installation, and so forth. (Much of this was implicit knowledge, going well beyond that officially taught in trade school.) If vocational trades education were to mimic this entire process, albeit as an educational exercise without the real constraints of workplace time pressures and power relations but within a given budget, I believe it would make a large contribution towards overcoming the problems of lack of meaning in mathematics. However, great care must be taken not to assume that mathematical skills and concepts are well understood by learners — without the underpinning knowledge, they will be unable to move beyond the routine and to create new knowledge to meet new problematic situations. In terms of the Kuutti model, communication between people would play a vital role unlike the case with most regular textbook or worksheet mathematics exercises. Also, the motive for learning would be much more apparent to learners, as they experience contradictions between what they would like to be able to do and their current levels of development, as they move up and down the hierarchies between operations, actions, and activity.

Of course, it may be argued that what I have proposed above is an ideal situation, and that adult numeracy and vocational mathematics curriculum is already prescribed and the classes have been timetabled separately. In this case, my experience with working in a pharmaceutical manufacturing workplace suggests that the prescribed learning outcomes can be addressed by the teacher concerned spending time carefully observing the activities relevant to the particular group of learners (operators in my case). Having the confidence to work without the ubiquitous textbook or photocopied worksheets, I was able to transform a lacklustre 'basic mathematics' course focused on recording and calculating with whole numbers, decimals and fractions to design learning opportunities which integrated the work that these people were already doing – and very competently so. Fortunately, at the same time I was also teaching — rather, 'delivering' — 'basic computing' to the workers, some of whom were much more competent than myself, and this gave me the opportunity to enhance the numeracy component by incorporating an industrial statistics package I was using with much higher level vocational students (see FitzSimons 2000, 2001, for more details).

In terms of activity theory, the workers were initially surprised to recognise the quality control [QC] charts they had often seen used by supervisors in production meetings. Our activity enabled them to be able to construct their own on the computer as a summary of their concrete dice-rolling experiments. Having been aware of QC charts in the abstract they were able to understand them concretely as a result of their own actions. The standard software producing the QC charts is an example of a reified cultural tool based upon the historical method of experimenting, then observing, recording, summarising, and displaying information. In addition to using authentic workplace contextualisations it is possible to draw on the life experiences of the particular learners or even to extend the possibilities. For example, it is reasonable to include topics of general adult interest such as the mathematics of gambling or current items of interest in the popular media in order to address the predetermined criteria in ways that are meaningful to the learners. As part of the focus on decimals I included a visit to the university library where the workers — some of whom had never been in a library — were able to locate items of personal interest using the Dewey decimal cataloguing system.

The Teaching and Learning Research Project team from the Institute of Education, London University, has conducted extensive work on numeracy — or *techno-mathematical literacies*, in their words — in the workplace and they strongly support the need for contextualisation. As Bakker et al. (2006) note:

> The special property of mathematical symbols is that in order to deal with them, to infuse them with meaning, it is necessary to situate them internally as part of a relationship between signs as well as externally, as part of a more or less familiar relationship between things (some of which might, of course, be symbolic). The balance between this internal and external system of relationships is one of the ways to recognize 'mathematical' activity of a professional kind. (p. 359)

Clearly, teaching for meaning is critical for adult numeracy and vocational mathematics learners, so that the problems associated with 'transmission-plus-application' teaching no longer arise. Activity theory provides an holistic framework on which to develop pedagogical materials that encompass the full range of activities from a socio-cultural perspective. In the next chapter I will focus on planning and evaluation of formal programmes, but it also possible — as Engeström (2001) shows — to describe and analyse an informal learning situation in order to resolve the tensions and contradictions which have arisen and to move on through an expansive cycle.

*Chapter 7*

# A FRAMEWORK FOR PLANNING AND EVALUATING FORMAL MATHEMATICS AND NUMERACY LIFELONG LEARNING PROGRAMMES

In order to plan methodically for a programme which takes into account the range of issues described in this book, and then to evaluate the outcomes against the intentions, the following grid devised by Engeström (2001) can serve a useful purpose through its prompting of useful analytical questions to assist in the process. Whereas planning is focused on intended outcomes, evaluation encompasses both intended and unintended consequences, with the opportunity for modifications to the programme based upon consideration of those consequences to encourage positive aspects and avoid where possible or ameliorate the negative ones.

Engeström (2001) outlined a framework for *expansive learning* — working towards a reconceptualisation of Activity theory. As discussed earlier, building on the work of Vygotsky he generated a structure for a human activity system, and then a model for two or more interacting activity systems, in order to "develop conceptual tools to understand dialogue, multiple perspectives, and networks of interacting activity systems" (Engeström, 2001, p. 135). He elaborated five principles to summarise Activity theory (discussed below) and cross-tabulated these with four questions central to any theory of learning — who are the subjects of learning, how, why, and what do they learn? (See Figure 2.) Mathematics or numeracy in adult and vocational

| | Activity system as a unit of analysis | Multi-voicedness | Historicity | Contradictions | Expansive cycles |
|---|---|---|---|---|---|
| Who are learning? | • Adults returning to study [post-16]<br>• Their teachers<br>• Community: employers, friends, family, etc. | • Adult learners' workplace, social, family responsibilities<br>• Teacher responsibilities | • Traditional division of labour | | |
| Why do they learn? | • Personal development & fulfilment<br>• Social, cultural, political etc. development<br>• Technological and socio-economic development. | • Supporting other family members' needs<br>• Improve career and/or further study choices<br>• Employer expectations<br>• Government requirements | • Pressures to engage in learning to avoid or cope with unemployment<br>• Programmes already institutionalised | • Between encouragement & coercion<br>• Between personal, social, & economic expectations<br>• Between personal needs & institutional offerings | Dialogue and debate recognising and reconciling adults' mathematical & numeracy learning needs |
| What do they learn? | • Mathematics<br>• Numeracy<br>• How to learn | • Explicit and tacit skills & knowledges | • Old and new forms of content | • Traditional skills may not be valued | Dialogue and debate on changing content |
| How do they learn? | • Formal<br>• Semi-formal<br>• Informal<br>• Face-to-face<br>• Online<br>• Community-based | • Possibly different kinds of learning at school and in workplace | • Old and new forms of pedagogy & delivery | • Prior learning expectations vs. pedagogical & technological innovations | Policy, teaching and learning actions to address transformative learning for adults |

Figure 2. Matrix for the analysis of expansive learning (following Engeström, 2001, p. 138).

education is complex and may be viewed and construed from many valid perspectives, according to the interests of the persons or groups concerned. In this context the concept of learner is taken broadly, to include *all* participants in the dynamic process — not just the students.

## ACTIVITY SYSTEM AS A UNIT OF ANALYSIS

Firstly, within Activity theory, the prime *unit of analysis* is taken to be "a collective, artefact-mediated and object-oriented system," subordinating goal-directed individual and group actions as well as automatic operations. "Activity systems realize and reproduce themselves by generating actions and operations" (Engeström, 2001, p. 136). Here, the system of interest could be the adult numeracy or vocational mathematics class comprising learners, their teacher/s (real or virtual), the material, communication, and thinking tools, set in the context of the community, rules, and division of labour. There are various, sometimes conflicting personal and societal motives and the kinds of learning may be formal, semi-formal (e.g., being shown how to do something in the workplace), or informal (the kinds of learning continually taking place throughout life).

## MULTI-VOICEDNESS

Secondly, account needs to be taken of the principle of *multi-voicedness*. The community within an activity system will have multiple points of view, traditions, and interests, and the participants will bring their individual and diverse histories. Different positions for participants are created by the division of labour within the activity system and the system itself carries its own history, reflected in its artefacts, rules, and conventions. This multi-voicedness can be a source of conflict and also innovation.

In this case multi-voicedness could include the multiple views, traditions and interests of learners, teachers, managers, employers, and government bodies. What are the reasons for individual learners for enrolling in the formal mathematics or numeracy programme? How are these needs to be taken into consideration? What are the unique histories of experience with mathematics and with mathematics education of both students and teachers as they enter this learning arena? What is known of their life and work experiences? Is it assumed that teachers should indeed learn about the needs of their students and their relevant industries or communities? Governments, through their bureaucratic processes, need to be able to guarantee some form of public accountability for funding spent on adult and vocational education. How much does this impact on the programme and in what ways? From a pedagogic perspective, is there — or ought there to be — dialogue and debate between positions and voices, focused (in this case) on mathematics and numeracy education? Overall, how — if at all — is the programme directed to the local and/or global economic, social, and environmental spheres?

## HISTORICITY

Given that activity systems evolve and become transformed over time, in order to understand the problems and potentials, *historicity*, the third of Engeström's (2001) principles is "studied as local history of the activity and as history of the theoretical ideas and tools that have shaped the activity" (pp. 136-137).

What is the background of the particular programme? When did it originate? Why? And for whom? Are these reasons still valid? Where is it located? What, if any, is the role of technology as a mode of delivery? Are there pressures for change in any of these? What is the impact of government policies — for example, economic rationalism, or policies supporting lifelong education, learning for work, or work and study for financial support, and so forth, on the conditions of teaching and learning? What financial support is available for material resources, including technology? Are the teachers adequately qualified? Is there sufficient administrative support? Is the timetable appropriate to all learners?

What are the philosophical bases for the content and the pedagogy? To the extent that they implicate mathematics, policies are underpinned by certain, historically constituted, understandings of what mathematics is, to whom mathematical knowledge should be distributed, and how much. What is at issue are historicised, necessarily partial, accounts of: (a) mathematics as a discipline, (b) the role of mathematics in vocations and society generally, and (c) the crucial role of mathematics education in achieving desired social and economic outcomes. Are the programme content, pedagogy, and form/s of delivery appropriate to all learners in quantity and quality?

## CONTRADICTIONS

Engeström's (2001) fourth principle claims a central role for *contradictions*. These are a source of change and development, and may be considered as an indication of historically accumulating structural tensions within and between activity systems. The primary contribution is between use and exchange value — and this looms large in formal mathematics education. Is the learning intrinsically useful and to be valued for its own sake or for its exchange value in the opportunities afforded by the qualification? Engeström raises an example of an activity system adopting a new element from outside, such as a new technology, leading to an aggravated secondary contradiction where old elements such as the rules or division of labour are no longer appropriate. Such contradictions may generate disturbances and conflicts, but may also lead to innovation, according to Engeström.

In terms of pedagogical practice of adult numeracy and vocational education in mathematics there are many contradictions: for example, between intended curricula and actual workplace and community practice, between teaching resources available (including adequately trained staff and well supported new technologies) and intended outcomes, and so forth. What are the contradictions faced by practitioners and students in particular adult numeracy and vocational mathematics education programmes?

Given that students are likely to have agency in various other arenas of life — as workers, parents, members of the community, and so forth — on entering the adult numeracy or vocational mathematics education programme, is it possible that they become (once again, as in school days) positioned as relative inferiors? Is there resistance by those who perceive themselves competent in life without the need for formal study of mathematics (Wedege, 1999)?

Are teachers forced to work under situations of overt or implicit control via mechanisms such as mandated competency-based training or International Standards Organisation (ISO) criteria, for example? Are there conflicts between what teachers' experience suggests and what their institutions demand by way of 'public accountability' (FitzSimons, 2002)? Are assessment-driven processes of moderation, while offering novices a useful form of professional development, problematic in their subtle forms of control over more experienced teachers?

Is the government of the day prone to projecting contradictory messages such as echoing the calls from business and industry for graduates who are creative problem-solvers, while simultaneously instituting curricula designed to encourage conformity and docility (e.g., FitzSimons, 2002)? Has the rhetoric of lifelong learning (e.g., Delors, 1996) as a social good apparently been adopted but economic imperative is made the overriding priority?

Related to power, in diverse and complex ways the concept of technology is central to mathematics in adult numeracy and vocational education. Firstly, from an industrial perspective it is integrally linked with mathematics in production (in the sense of the totality of effort, material, intellectual, symbolic, etc.) in manufacturing, service, and symbolic-analytic sectors. How well are the industrial or vocational uses of technology integrated into the particular programme? Secondly, technology is utilised as a tool of management control in industrial and vocational settings. Are the students enrolled in the programme enabled to adopt a mathematically critical attitude to understand and, where necessary, challenge inappropriate uses of mathematical and statistical tables, charts, diagrams, and so forth? Thirdly, does technology as a medium of instruction in an increasingly a competitive education industry (FitzSimons, 2007), within individual classrooms and across national or international sites of learning, take into account the multi-voicedness and the historicity aspects described above, and within an ethical framework?

## EXPANSIVE LEARNING

Engeström's (2001) fifth principle proclaims the possibility of *expansive cycles* — "expansive transformations in activity systems" (p. 137). In these relatively long cycles of qualitative transformations, as contradictions become aggravated questioning and deviation from established norms sometimes escalates into a deliberate collective change effort. The object and motive are reconceptualised to embrace a radically wider horizon of possibilities. According to Engeström (1999, p. 137) "a full cycle of expansive transformation may be understood as a collaborative journey through *zone of proximal development* [ZPD] of the activity." As mentioned earlier, Engeström also notes that, contrary to tradition notions of stable and reasonably well defined identifiable knowledge or skills and a competent teacher, learning in work organisations violates these presuppositions in that the objects of most learning are not stable nor even defined or understood ahead of time. Transformational learning requires new forms of activity that are literally learned as they are created and for which there is no competent teacher.

Engeström (2001, p.153) suggests that a complementary perspective to the traditional view of learning and development as vertical processes be constructed: horizontal or sideways development. Instead of trying to merge possibly incompatible scientific concepts with everyday experiences, he recognised that a series of alternative conceptualisations could be useful in resolving tensions and contradictions. In the continuing process of expansive learning the workers in Engeström's study were able accumulate experiences to further challenge and transform their newly elaborated, refined, and concretised solution concept. In my opinion there is a resonance between Engeström's conceptualisation of horizontal development and Bernstein's concept of horizontal discourse as I have applied them to mathematics and numeracy practice and education. Engeström's model of horizontal development could be used to reconcile other contradictions that arise in education — such as those identified above — through alternative conceptions based on respectful dialogue between the stakeholders.

Is it possible that Critical Mathematics (Skovsmose, 1994) may be a serious part of the intended and implemented curriculum for adult and vocational education? What might be a role for aesthetics (e.g., Bagnall, 1997)? How might it be possible to value both cognitive and affective development in planning and evaluation? How might alternative, research-based, curricular and pedagogical practices be incorporated into adult numeracy and vocational mathematics? How might alternative, more positive, public perceptions of mathematics and mathematics education be generated and enhanced to encourage the uptake of lifelong learning in ways that benefit all stakeholders? More broadly, might it be possible to involve some or all of the stakeholder groups in a collective change effort to serve the personal, social, economic, and environmental needs of our planet through more appropriate mathematics and numeracy education?

*Chapter 8*

# CONCLUSION

The following recommendations emerged from the UNESCO/UIL (2008) meeting in Korea:

- The concept of lifelong learning should be concretised and elaborated;
- The lifelong learning discourse should be linked to other educational discourses;
- There is a need for advocacy for a lifelong learning framework as well as a legislative and administrative framework;
- Formal, non-formal and informal learning should be integrated;
- Good practices in lifelong learning policies should be shared. (p. 26)

In this book I have attempted to address each of these, to different degrees, from my own perspective as a practitioner and researcher. The UNESCO (2005) publication, *Towards Knowledge Societies*, echoes the sentiments about lifelong learning expressed in the Introduction to this book, underlining the notion that lifelong learning is a process that should ideally be meaningful at three, closely connected, levels: personal and cultural development, social development, and professional development. This is no less true for mathematics and numeracy education than for any other area of lifelong learning. The report recognises the danger of adults becoming permanently entrapped in an expanding vocational education and training industry with the risk of learners' needs being subordinated to those of industry. Instead, it recognises that "citizens must be able to express their own educational aspirations and choices" (p. 80). It recognises that not all people are equipped with the skills to manage and organise their own long-term learning paths, and that there may be conflicts for learners such as the amount of time they spend in training as opposed to work. It also raises the concerns of older learners who may be feeling that their previous learning is no longer valued, and that new ways of working, and forms of assessment are now required. These ideas reinforce the importance of an activity theory approach to understanding the social, cultural, historical, and economic contexts together with the tensions and contradictions faced by lifelong learners.

More optimistically the UNESCO (2005) report suggests

> ... placing individuals in an educational continuum in which their knowledge, skills and outlook on the world is continually renewed and consolidated, might provide them with the benefits of advances in new technologies and, above all, the implementation of efficient and diversified systems of education. (p. 81)

Clearly technology is having and will continue to have a major role in education in those parts of the world where access is available and reliable. More than just technical skills, the UNESCO (2005) report stresses that it is important for adults to learn how to "*choose* from the increasingly abundant array of educational and other software and educational programmes on offer, those that are most appropriate" (p. 83). It is apparent that the internet is rapidly becoming the most important medium of self-instruction through its provision of tools for informal learning as well as allowing for the creation of virtual classes. Discussing the innovative Virtual High School project in the USA, two major difficulties were noted: (a) the lack of educational policies on online teaching in many states, and (b) the absence of quality standards in online education which designers need to meet. Other challenges to the successful implementation of technology-based education are that virtual education requires substantial support in terms of material infrastructures of major equipment, high speed web connections, supported by competent engineers and net administrators. That is, installation of sophisticated equipment is not sufficient if the necessary timely maintenance is not available — as can happen in remote areas such as rural and outback Australia. The report also identifies another important consideration in the Virtual High School project: It is based upon complementarity between distance education and face-to-face teaching and not on substitution. In fact the UNESCO report calls for the relationship between distance and learning to be clarified, noting that communication is more than retrieving and exchanging information electronically to build up and share knowledge. It is important that the sharing relationship inherent in teaching and learning not be diminished or lost as a result of the new and diverse ways of acquiring knowledge. In other words, while high levels of technical support are necessary, education and learning are essentially people-centred activities.

Following their interests, locally and globally, adults will learn when they see the need for so doing, even when such learning presents challenges, cognitively and emotionally — particularly in the case of mathematics. Sociocultural historic activity theory has offered a framework for analysis of technology-supported learning in the realm of operations, actions, and activities that a group of learners may participate in. Engeström's grid has also offered a framework for the planning and evaluation of any programme for adult learners that takes into account who they are, what they bring with them as citizens, workers, and so forth, their education backgrounds, why they are learning in the current situation, how they learn, and what they intend to learn. All of these are set in the context of the history of the programme as a whole, the artefacts used together with their histories, the broader community involvement, and the division of labour. Most usefully, perhaps, activity theory alerts us to the possible tensions and contradictions which are the source of new learning, pointing the way to expansive learning in order to resolve problematic situations and meet new challenges as they arise.

# REFERENCES

Aspin, D. & Chapman, J. (2001). Towards a philosophy of lifelong learning. In D. Aspin, J. Chapman, M. Hatton, & Y. Sawano (Eds.), *International handbook of lifelong learning* (pp. 3-33). Dordrecht/Boston/London: Kluwer Academic Publishers.

Bagnall, R. G. (1997). Aesthetic knowledge and learning in the workplace. In *Good thinking — Good practice: Research perspectives on learning and work. Proceedings of the 5th Annual International Conference on Post-Compulsory Education and Training* (Vol. 1), (pp. 13-26). Brisbane, Qld: Centre for Learning and Work Research, Griffith University.

Bakker, A., Hoyles, C., Kent, P. & Noss, R. (2006). Improving work processes by making the invisible visible. *Journal of Education and Work, 19*(4), 343-361.

Borba, M. C. & Villarreal, M. E. (with D'Ambrosio, U. & Skovsmose, O.). (2005). *Humans-with-media and the reorganization of mathematical thinking: Information and communication technologies, modeling, visualization and experimentation.* NY: Springer.

Civil, M. & Quintos, B. (2009). Latina mothers' perceptions about the teaching and learning of mathematics: Implications for parental participation. In B. Greer, S. Mukhopadhyay, A.B. Powell, & S. Nelson-Barber (Eds.), *Culturally responsive mathematics education* (pp. 321-343). New York & London: Routledge.

Cockcroft, W. H. (Chairman) (1982). *Mathematics counts: Report of the Committee of Inquiry into the Teaching of Mathematics in Schools.* London: Her Majesty's Stationery Office.

Cohen, D. (Ed.) (1995). *Crossroads in mathematics: Standards for introductory college mathematics before calculus.* Memphis, TN:

American Mathematical Association of Two-Year Colleges [AMATYC]. [Retrieved November 7, 2008 from the World Wide Web: http://www.imacc.org/standards/]

Cranton, P. A. (1992). *Working with adult learners.* Toronto: Wall & Emerson.

Cranton, P. A. (2006). *Understanding and promoting transformative learning: A guide for educators of adults* (2nd ed.). San Francisco, CA: Jossey-Bass.

Daly, R. & Mjelde, L. (2000). Learning at the point of production: New challenges in the social divisions of knowledge. In D. Boud & C. Symes (Eds.), *Working knowledge: Productive learning at work. International conference proceedings* (pp. 105-113). Sydney: Research into Adult and Vocational Learning, University of Technology, Sydney.

Delors, J. (Chair). (1996). *Learning: The treasure within.* Report to UNESCO of the International Commission on Education for the Twenty-first Century. Paris: United Nations Scientific, Cultural and Scientific Organization (UNESCO).

Díez-Palomar, J. (2007). Una aproximación dialógica de la inclusión en matemáticas en la escuela obligatoria: el caso del razonamiento proporcional. In J. Giménez, J. Díez-Palomar, M. Civil (Eds.), *Educación matemática y exclusión* (pp. 147-178). Barcelona: Editorial GRAÓ.

Engeström, Y. (1987). *Learning by expanding: An activity-theoretical approach to developmental research.* Helsinki: Orienta-Konsultit.

Engeström, Y. (2001). Expansive learning at work: Toward an activity-theoretical reconceptualization. *Journal of Education and Work, 14*(1), 133-156.

Ernest, P. (1991). *The philosophy of mathematics education.* Hampshire, UK: Falmer Press.

European Union (1996a). *Living and working in the information society: People first.* (Green paper.) Retrieved 5 February, 2001, from the World Wide Web: http://europa.eu.int/ISPO/infosoc/legreg/docs/people1st.html

European Union (1996b). *Vocational Training and Education: European Year of Lifelong Learning: 1996.* Retrieved 16 November, 2000, from the World Wide Web: http://europa.eu.int/scadplus/leg/en/cha/c11024.htm

European Union (2000). *A memorandum on lifelong learning. Commission staff working paper.* Brussels: Commission of the European Communities. Retrieved 5 February, 2001, from the World Wide Web: http://europa.eu.int/comm/education/life/memoen.pdf

FitzSimons, G. E. (2000). Lifelong learning: Practice and possibility in the pharmaceutical manufacturing industry. *Education & Training, 42(3)*, 170-181.

FitzSimons, G. E. (2001). Integrating mathematics, statistics, and technology in vocational and workplace education. *International Journal of Mathematical Education in Science and Technology, 32(3)*, 375-383.

FitzSimons, G. E. (2002). *What counts as mathematics? Technologies of power in adult and vocational education.* Dordrecht: Kluwer Academic Publishers.

FitzSimons, G. E. (2003). Using Engeström's expansive learning framework to analyse a case study in adult mathematics education. *Literacy & Numeracy Studies, 12(2)*, 47-63.

FitzSimons, G. E. (2007). Globalisation, technology, and the adult learner of mathematics. In B. Atweh. A. Calabrese Barton, M. Borba, N. Gough, C. Keitel, C. Vistro-Yu, & R. Vithal (Eds.), *Internationalisation and globalisation in mathematics and science education* (pp. 343-361). Dordrecht: Springer.

FitzSimons, G. E. (2008a). A comparison of mathematics, numeracy and functional mathematics: What do they mean for adult numeracy practitioners? *Adult Learning Quarterly*.

FitzSimons, G. E. (2008b). Mathematics and numeracy: Divergence and convergence in education and work. In C. H. Jørgensen, & V. Aakrog (Eds.) *Convergence and Divergence in Education and Work. Studies in Vocational and Continuing Education series, vol. 6* (pp. 197-217). Zurich: Peter Lang AG.

FitzSimons, G. E., & Wedege, T. (2007). Developing numeracy in the workplace. *Nordic Studies in Mathematics Education, 12(1)*, 49-66.

Freire, P. (1972). *Pedagogy of the oppressed* (M. Bergman Ramos, Trans.). Harmondsworth: Penguin Books.

Hedegaard, M., Chaiklin, S., & Jensen, U. J. (1999). Activity theory and social practice: An introduction. In S. Chaiklin, M. Hedegaard, & U. J. Jensen (Eds.), *Activity theory and social practice* (pp. 12-30). Aarhus: Aarhus University Press.

Hoyles, C., Wolf, A., Molyneux-Hodson, S. & Kent, P. (2002). *Mathematical skills in the workplace. Final report to the Science, Technology and Mathematics Council.* London: Institute of Education, University of London: Science, Technology and Mathematics Council.

Knowles, M. S. (1990). *The adult learner: A neglected species* (4th ed.). Houston, TX: Gulf Publishing Company.

Kolb, D. A. (1984). *Experiential learning*. NJ: Prentice-Hall.
Kuutti, K. (1996). Activity theory as a potential framework for human-computer interaction research. In B. A. Nardi (Ed.), *Context and consciousness: Activity theory and human-computer interaction* (pp. 17-44). Cambridge, MA: The MIT Press.
Lave, J. (1988). *Cognition in practice: Mind, mathematics and culture in every day life*. Cambridge: Cambridge University Press.
Maglen, L. (1996). *VET and the university*. Inaugural professorial lecture. Melbourne: The University of Melbourne.
Mezirow, J. (1996). Contemporary paradigms of learning. *Adult Education Quarterly, 46*(3), 158-173.
National Council of Teachers of Mathematics. (2000). *Principles and standards for school mathematics*. Reston, VA: Author.
Niss, M. (1996). Goals of mathematics teaching. In A. J. Bishop, K. Clements, C. Keitel, J. Kilpatrick, & C. Laborde (Eds.), *International handbook of mathematics education* (pp. 11-47). Dordrecht: Kluwer Academic Publishers.
Organisation for Economic Co-operation and Development [OECD]. (1996). *Lifelong learning for all*. Paris: Author.
Rogers. C. (1969). *Freedom to learn*. Columbus, OH: Charles E. Merrill.
Skovsmose, O. (1994). *Towards a philosophy of critical mathematics education*. Dordrecht: Kluwer Academic Publishers.
Stetsenko, A. P. (1999). Social interaction, cultural tools and the zone of proximal development: In search of a synthesis. In S. Chaiklin, M. Hedegaard, & U. J. Jensen (Eds.), *Activity theory and social practice* (pp. 235-252). Aarhus: Aarhus University Press.
Tennyson, R. D., & Schott, F. (1997). Instructional design theory, research, and models. In R. Tennyson, F. Schott, N. Steel, & S. Dijkstra (Eds.) *Instructional design: International perspectives* (Vol. 1: Theory, research, and models), (pp. 1-16). Mahwah, NJ: L. Erlbaum Associates.
Trouche, L. (2004). Managing the complexity of human/machine interactions in computerized learning environments: Guiding students' command process through instrumental orchestrations. *International Journal of Computers for Mathematics Learning, 9*(3), 281-307.
UNESCO. (2000). *World education report 2000: The right to education: Towards education for all throughout life*. Paris: UNESCO Publishing.
UNESCO & International Labour Organisation (ILO) (2002). *Technical and vocational education and training for the twenty-first century: UNESCO and ILO recommendations*. Paris: UNESCO Publishing.

UNESCO (2005). *Towards knowledge societies*. Paris: UNESCO Publishing. [Available from: http://unesdoc.unesco.org/images/0014/001418/141843e.pdf]

UNESCO Institute for Lifelong Learning (UIL) (2008). *Annual report, 2007*. Paris: UNESCO Publishing. [Available from: http://unesdoc.unesco.org/images/ 0015/001598/ 59840m.pdf]

Wedege, T. (1999). To know — or not to know — mathematics, that is a question of context. *Educational Studies in Mathematics, 39(1-3)*, 205-227.

# INDEX

## A

accommodation, 13
accountability, 8, 46, 48
accuracy, 8
achievement, 7, 8, 18
activity level, 35
activity theory, 3, 21, 24, 28, 41, 52, 53
administrative, 46, 51
administrators, 52
adult, xvii, 3, 4, 6, 7, 8, 9, 10, 11, 12, 13, 14, 16, 19, 20, 25, 26, 31, 34, 37, 38, 39, 40, 41, 45, 46, 47, 48, 49, 53, 56, 57
adult education, 4, 10, 11, 12, 19, 20
adult learning, 4, 39
adults, xvii, 3, 4, 6, 8, 12, 14, 16, 19, 20, 25, 31, 32, 33, 38, 44, 51, 52, 53, 56
advocacy, 51
aesthetics, 49
age, 2
alienation, 10
alternative, 49
anxiety, 8
arithmetic, 26, 32, 38
armed forces, 8
Asia, vii
assessment, 7, 11, 13, 19, 48, 52
assumptions, 6, 12

attitudes, 4, 20
Australia, 8, 9, 10, 11, 14, 52
Australian Research Council, 31
authoritarianism, 8
authority, 5
automation, 34, 35
autonomy, 9
avoidance, 8
awareness, 27

## B

back, 8
barriers, 9
beliefs, 1, 18
benefits, 52
blocks, 33
Boston, 55
Brazil, 35
Britain, xvii
Brussels, 56
buildings, 25

## C

calculus, 32, 55
campaigns, 19
case study, 57
certification, 8
children, 9, 11, 16, 19, 25, 38, 39

China, ii, viii
citizens, 20, 51, 53
citizenship, 13, 16
classes, 3, 8, 10, 11, 20, 38, 40, 52
classroom, xvii, 7, 17, 19, 27, 28, 39
classrooms, 8, 26, 27, 48
coercion, 4, 44
cognitive level, 12
cognitive psychology, 24
collaboration, 24, 28, 35
commerce, 37
communication, xviii, 20, 24, 28, 33, 40, 45, 52, 55
communication technologies, 28, 55
communities, 9, 13, 16, 34, 46
community, 2, 3, 25, 26, 28, 32, 33, 45, 47, 53
competence, 10, 18
competency, 39, 48
competition, 2
complementarity, 52
complex systems, 24
complexity, 16, 25, 28, 58
components, 18, 26, 31
compulsory education, 20
computer software, 10, 20
computing, 41
conceptualization, 11
concrete, 10, 11, 18, 20, 26, 41
confidence, 10, 11, 14, 40
conflict, 13, 46
conformity, 48
Congress, xiv
consciousness, 58
constraints, 13, 14, 25, 40
construction, 1, 17, 19, 34
constructivist, 9, 10, 13
context-dependent, 19
contracts, 9
control, 5, 8, 48
convergence, xvii, 57
costs, 40
creative thinking, 27
creativity, 35
credentials, 15

criticism, 1
cultural practices, 38
cultural values, 23
culture, 58
curiosity, 1
curriculum, 9, 10, 37, 38, 40, 49
cycles, 44, 48

## D

danger, 51
data collection, 20
decision makers, 3
decision making, 13
decisions, 8, 11, 16
definition, 17, 38
delivery, 5, 7, 13, 28, 44, 46, 47
designers, 52
developed countries, 4
developing countries, 34
deviation, 48
differential rates, 38
disability, 33
discipline, 3, 5, 6, 10, 15, 16, 17, 18, 19, 25, 28, 47
discourse, 14, 16, 17, 18, 19, 27, 38, 40, 49, 51
discrimination, 39
disorder, 27
distance education, 52
diversity, 13, 25
division, 3, 16, 17, 25, 26, 28, 32, 33, 34, 44, 45, 46, 47, 53
dominance, 19
duties, 14

## E

ears, 17
echoing, 48
economic development, 2, 44
Education, i, ii, iii, iv, v, vi, vii, viii, ix, x, xiii, xv, 3, 5, 9, 19, 31, 41, 55, 56, 57, 58

educational policies, 52
educational settings, 34
educators, 11, 12, 14, 56
email, 28, 33
emotional, 12
emotions, 18
employees, 2, 33
employers, 15, 38, 39, 44, 46
employment, 25
empowerment, 11
encouragement, 44
enculturation, 16, 18
English Language, viii
environment, 17
epistemological, 1
European Union, 56
examinations, 6, 7
exercise, 40
expertise, 10, 19
explicit knowledge, 18

## F

failure, 9
family, 11, 16, 44
family members, 44
fears, 14
February, 56
feedback, 32
females, 11
financial support, 46
formal education, 3, 4, 14, 16, 20, 26, 28, 37
frustration, 6, 8
funding, 46

## G

gambling, 41
games, 10
gender, 7
goal-directed, 31, 45
goals, 3, 17, 18, 26, 27
government, 3, 9, 16, 46, 48

grades, 39
group work, 9
groups, 9, 16, 26, 33, 45, 49
growth, 9

## H

heart, 23
heating, 40
higher education, 14, 39
high-level, 18
holistic, 13, 40, 42
Holland, 11
horizon, 49
human, xviii, 2, 23, 24, 37, 45, 58
human activity, 45
human development, 24
humans, 23, 35
humiliation, 6, 8, 37
hybrid, 13

## I

ILO, 1, 58
images, 59
imagination, 13
immigrants, 26, 34
implementation, 3, 52
implicit knowledge, 40
in situ, 37
incarceration, 33
indeterminacy, 27
indication, 47
industrial, 20, 41, 48
industry, 39, 48, 51, 57
information and communication technologies, 28
injury, xiv
innovation, 46, 47
institutions, 1, 2, 4, 19, 48
instruction, 20, 28, 33, 48, 52
instructional design, 4, 32
instruments, 25
integration, 17

intelligence, 15, 37
intentions, 43
interaction, 5, 23, 24, 35, 58
interactions, 25, 58
International Labour Organisation, 58
internet, 6, 20, 28, 52
interrelationships, 16
ISO, 48
isolation, 8, 39

## J

jobs, 13
justice, 13
justification, 11

## K

Korea, 51

## L

labour, 2, 3, 25, 26, 28, 32, 33, 34, 37, 44, 45, 46, 47, 53
labour-saving, 26, 28, 37
language, 6, 17, 38
law, 23
learners, 1, 3, 5, 6, 7, 8, 9, 10, 11, 12, 13, 16, 19, 20, 25, 26, 27, 28, 31, 32, 33, 34, 35, 38, 39, 40, 41, 44, 45, 46, 47, 51, 53, 56
learning, xvii, 1, 2, 3, 4, 6, 7, 8, 9, 10, 11, 12, 13, 14, 16, 19, 20, 21, 24, 25, 26, 27, 28, 29, 31, 33, 34, 35, 37, 38, 39, 40, 42, 44, 45, 46, 47, 48, 49, 51, 52, 53, 55, 56, 57, 58
learning activity, 24, 25, 28
learning culture, 26
learning environment, 58
learning outcomes, 13, 19, 40
learning process, 11, 12, 24, 28
learning task, 27
leg, 56
lens, xvii

liberation, 12
life experiences, 41
lifelong learning, xvii, 1, 2, 4, 9, 14, 20, 26, 28, 35, 39, 48, 49, 51, 55, 56
lifetime, 16, 18
limitations, xviii, 34, 37
links, 17
literacy, xvii, 19
localised, 19
location, 20, 28, 33
locus, 38
London, 41, 55, 57
love, 6
low-level, 31

## M

machinery, 20
magnetic, xiv
mainstream, 2, 19, 26, 33
mainstream society, 33
maintenance, 2, 52
management, 2, 13, 14, 48
manipulation, 32
manufacturing, 13, 40, 48, 57
markets, 39
mastery, 6, 7, 8, 18, 25, 27, 32
material resources, 46
mathematical knowledge, 17, 18, 47
mathematical skills, 7, 15, 16, 40
mathematical thinking, 55
mathematicians, 9, 10, 16, 38
mathematics, xvii, 2, 3, 4, 5, 7, 8, 9, 10, 11, 13, 14, 15, 16, 17, 18, 19, 20, 21, 25, 26, 28, 31, 32, 33, 34, 35, 37, 38, 39, 40, 41, 45, 46, 47, 48, 49, 51, 53, 55, 56, 57, 58, 59
mathematics education, xvii, 4, 8, 9, 15, 16, 17, 26, 31, 32, 34, 37, 39, 46, 47, 49, 55, 56, 57, 58
measurement, 32, 39
media, 15, 41, 55
memory, xviii, 37
messages, 48
metacognition, 27

metacognitive skills, 27
metaphor, 27
metric, 25, 32, 39
Millennium, ix
misleading, 16
MIT, 58
modalities, 26
modeling, 55
models, 10, 11, 18, 19, 58
mothers, 11, 55
motivation, 6, 7, 9, 14, 27
motives, 25, 45
movement, 8, 10, 11
multicultural, 10
multimedia, 35
multiplication, 16

## N

natural, 39
negative consequences, 3
negative experiences, 4
negotiation, 13, 34
neoliberal, 9
network, 33, 34
New York, xiii, xiv, 55
next generation, 23
normal, 2
norms, 25, 33, 48
nursing, 8

## O

obedience, 8
observations, 3, 11, 26, 28, 40
obsolete, 27
OECD, 1, 20, 58
offshore, 33
online, 7, 13, 20, 31, 33, 52
oppression, 12
Organisation for Economic Co-operation and Development, 58

## P

parental participation, 55
parents, 13, 19, 47
Paris, 56, 58, 59
pathways, 14
pedagogical, 20, 26, 42, 44, 47, 49
pedagogy, xvii, 5, 7, 13, 18, 19, 44, 47
peer, 24
perceptions, 3, 15, 49, 55
personal relations, 18
pharmaceutical, 13, 40, 57
philosophical, 1, 4, 47
philosophy, 13, 15, 55, 56, 58
planning, 3, 11, 40, 42, 43, 49, 53
Plato, 5
play, 12, 24, 39, 40
police, 8
policy makers, 7
politicians, 3
poor, 27
population, 15
power, 40, 48, 57
power relations, 40
practical knowledge, 19
press, 37
pressure, 13
printing, 37
problem solving, 9, 10, 27, 40
problem-solver, 48
production, 25, 26, 38, 39, 41, 48, 56
productivity, 14
professional development, 48, 51
program, 10, 33, 39
prototype, 23
proxy, 15, 37
pseudo, 7, 19, 32
public, 15, 16, 46, 48, 49
public education, 16
pupils, 6

## Q

qualifications, 2, 18, 39

quality control, 41
questioning, 13, 28, 48

## R

randomness, 38
range, 2, 3, 10, 18, 26, 42, 43
rationalist, 9
reading, 18
*Reading at Risk*, ii
reality, 27
reasoning, 28
recognition, 13
reconcile, 20, 25, 49
reconstruction, 26
reflection, 9, 10, 12, 34
regular, 27, 40
reinforcement, 32
relationship, 41, 52
relationships, 41
relevance, 2, 19
remediation, 32
reproduction, 25
resistance, 34, 47
resolution, 26
resources, 20, 25, 32, 46, 47
responsibilities, 28, 33, 44
responsibility for learning, 6, 8
rhetoric, 27, 48
risk, 51
rolling, 41
rote learning, 7
routines, 32, 35
rural, 52

## S

scaffolding, 27
school, xvii, 2, 3, 4, 6, 9, 10, 14, 15, 16, 17, 19, 26, 27, 37, 38, 40, 44, 47, 58
school work, 19
schooling, 2, 4, 9, 18, 25
science education, 57
scientific method, 9
search, 27, 32, 58
search engine, 32
searching, 6, 20, 28
Self, iv, 11
self-confidence, 8, 11
Self-Direction, 11
series, 49, 57
services, xiv, 8
shape, 20
sharing, 52
Shell, 10
shyness, 11
signs, 24, 41
simulation, xviii
sites, 2, 48
skills, 1, 6, 7, 15, 16, 20, 25, 26, 27, 28, 32, 37, 38, 39, 40, 44, 49, 51, 52, 57
social change, 12
social construct, 13, 16
social context, 9, 12, 20
social development, 51
social isolation, 8
social justice, 13
social life, 2, 25
social order, 12
social problems, 8
socialist, 13
software, xviii, 19, 28, 32, 34, 41, 52
spatial, 27, 39
species, 57
speed, 52
spheres, 2, 18, 46
spirituality, 13
spreadsheets, xviii, 28, 32
stages, 11
stakeholder, 49
stakeholder groups, 49
stakeholders, 16, 49
standards, 48, 52, 55, 56, 58
statistics, xviii, 19, 32, 34, 41, 57
strategies, xvii, 3, 10, 17, 18, 19
stress, 12, 27
students, 5, 7, 10, 11, 14, 20, 25, 26, 34, 37, 41, 45, 46, 47, 48, 58
Sub-Saharan Africa, iv

substitution, 52
subtraction, 16
supervisors, 41
surprise, 37
symbols, 24, 26, 41
synthesis, 24, 58

## T

tangible, 26
Taylorism, 7
teacher support, 10
teachers, 4, 6, 8, 10, 12, 20, 25, 26, 33, 38, 39, 44, 46, 48
teaching, xvii, 2, 3, 4, 5, 6, 7, 9, 10, 11, 13, 14, 16, 18, 19, 20, 27, 28, 33, 34, 38, 39, 41, 44, 46, 47, 52, 55, 58
teaching strategies, 10
temporal, 27
tension, xvii, 25
textbooks, 8, 10
thinking, xviii, 10, 25, 28, 45, 55
time pressure, 40
trade, 16, 39, 40
tradition, 6, 23, 24, 49
traditional model, 7
traditional views, 38
training, 8, 14, 48, 51, 58
transfer, xvii, 7, 17, 20, 39
transformation, 12, 13, 25, 49
transformations, 48
transmission, 5, 6, 7, 16, 18, 28, 41
transparent, 29, 33
two-dimensional, 31

## U

uncertainty, 27
unemployment, 8, 44
UNESCO, 1, 51, 52, 56, 58, 59
United Nations, 56
university education, 2

## V

vacuum, 27
values, 12, 23, 33
visible, 11, 33, 34, 55
visualization, 55
vocational, xvii, 2, 3, 4, 6, 14, 19, 26, 31, 37, 38, 39, 40, 41, 45, 46, 47, 48, 49, 51, 57, 58
vocational education, 2, 4, 19, 37, 45, 46, 47, 48, 49, 51, 57, 58
Vygotsky, 23, 45

## W

war, 10
wealth, 37
web, 31, 32, 52
women, 10
word processing, 28
workers, 2, 4, 13, 14, 20, 26, 41, 47, 49, 53
workforce, 2
workplace, xvii, 2, 4, 8, 13, 14, 16, 17, 19, 20, 26, 27, 28, 32, 34, 35, 40, 41, 44, 45, 47, 55, 57
World Wide Web, 56
writing, 24